The Mark Of Light On The Great Pyramid Of Giza

Addenda to '32.5 System'

Edition 1.9.2

IBRAHIM IBRAHIM

DEDICATION

To All Of Those Who Love Divine Proportions

CONTENTS

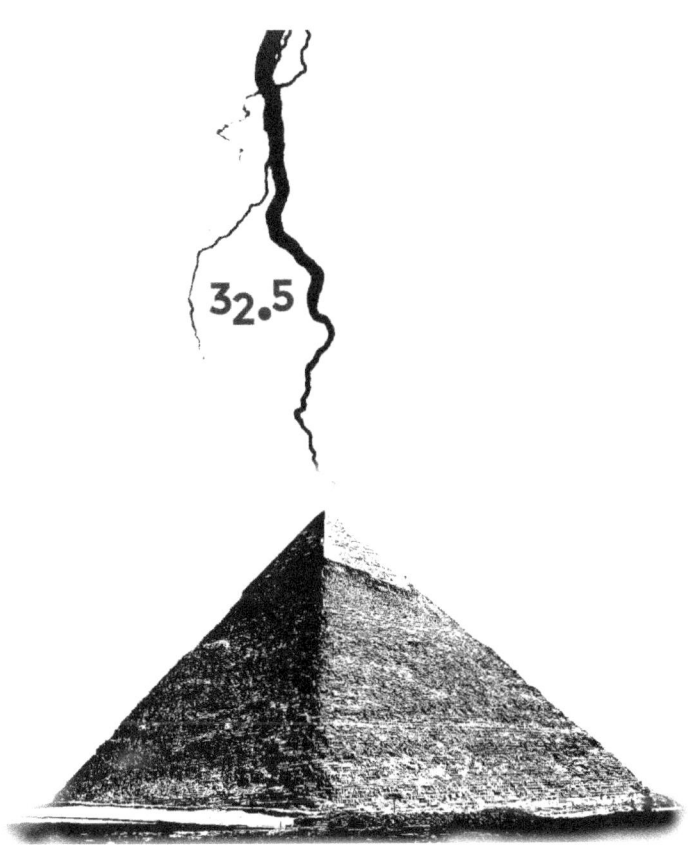

PREFACE

The research which is included in this book came to my surprise during the advanced studies I have been undertaking after about a year from publishing my book *'32.5 System'*. This is when I started observing some vivid *mathematical* relations in the structure of the Great Pyramid of Giza (GPG). The first spot took place when Robert Bauval embraced one of Gary Osborn's slides and presented it as relevant to the mysteries of the GPG. It has encouraged me enough to inspect Gary's slide and discovered thereupon that my work – which I presented in the book mentioned above – is part of that GPG's *narrative*!

I started examining the numbers that were incorporated into the GPG's *dimensions* and were laid down before my eyes on *facebook* until I realized that the amount of information I was able to generate alongside the discoveries I was able to make were big enough that they needed to get collected into a book.

Although the cover of the book does not directly (through text) reflect the material being presented and included therein, but it certainly does point to the theme behind the numbers themselves and also projects the path – beginning by 32.5 System – onto which this material was developed and the framework by which future studies will build up the ultimate narrative which I am articulating thereby.

I've received some inquires about why I have used Khafre's pyramid on the cover instead of Khufu's. Well, the reason is simple:

1. This is irrelevant to the topic being handled in the book since photography is not part of its topic.

2. The original shape of Khufu's pyramid cannot be seen nowadays on its structure, so the overall outer look is not that which matters, but rather its architectural design and hence the original numbers thereof.

3. Even though Khafre's pyramid has some tura casing blocks preserved on its top level courses, originally all three pyramids of the Giza Plateau were entirely covered with such casing blocks; hence, today's superficial view of Khafre's pyramid is not unique to it. For further information please visit:

   http://www.cheops-pyramide.ch/khufu-pyramid/casing-stones.html

4. Showing how the upper level courses look – as they are present on Khafre's pyramid on the cover, gives the reader a sense of relevancy to the contents of the book.

5. The book is not produced for tourism purposes of visual guidance.

6. On all of the images on the web of the Giza Plateau, the pyramid in the middle looks the biggest and gives the illusion as if it were that of Khufu's; but in reality it is rather that of Khafre's. The middle pyramid sits on a bedrock 10 m (33 ft) higher than Khufu's pyramid, which makes it appear to be taller. However, since there is no significant difference in viewing both pyramids, then the image on the cover can simply demonstrate how Khufu's pyramid looked like with tura casing.

The least we can say about the Great Pyramid of Giza is that it certainly was an observatory of the heavens.

INTRODUCTION

In this book, I document the various discoveries which I was able to embark upon and uncloak while inspecting the dimensions of the Great Pyramid of Giza (aka, Pyramid of Cheops). In addition to my own research of decoding the embedded numerical message of the Speed of Light value in this exceptional megalithic structure, I also present to the reader my interactions with world-renowned best-selling author, archaeoastronomer and Egyptologist: Robert Bauval.

The reader will have more information and insights herein of my claims to be the very first person who has:

    a. observed the significance of the number 32.5 in nature and in relation with the dimensions of this structure which is from the Seven Wonders of the Ancient World, i.e., the Pyramid of Giza.

    b. recognized that the size of King's Chamber in the Pyramid of Cheops is as important as its location inside the pyramidal complex.

and also the first person to discover:

    i. that the Pyramid's base area is that which tunes the proportions rather than its *perimeter*;

    ii. the link between the ancient Egyptian Royal Cubit -in reference to the metric System- with the Speed of Light (expressing this discovery numerically and textually);

iii.   the hidden factors behind having the difference between the Great Pyramid's two circles equal to the Speed of Light value and offered an explanation thereof;

iv.   the iterative design of the Great Pyramid's base diagonal which conserves the speed of light constant;

v.   the link between the Great Pyramid's dimensions and the horizon;

vi.   the Great Pyramid's height significance for serving as a direct connection to the Spring Equinox;

and much more.

Man was created of haste. I **will** show you My signs, so do not impatiently urge yourselves.

~ Quran 21:37

1 INTEREST IN THE GREAT PYRAMID OF GIZA

I hereby provide a timeline that includes a list of my discoveries in *chronological order* in the field of research in *Egyptology* which is concerned with examining the dimensions of the Great Pyramid of Giza (GPG). Therefore, before I advance my accomplishments in this *area* before the reader, I would like to clarify here in this chapter, the foundation onto which I have built my research and observations that resulted and culminated eventually into these findings.

I started my work and investigations in the field of alternative research back in 2012 after I watched the *documentary – Quest for the Lost Civilization* [1]; Starring: Graham Hancock and Robert Bauval. It didn't take long until the image of Mr. Hancock started crystalizing itself in my perception to resemble [T]he Journalist whose work I was yearning after from the days of my own childhood; having said that, I do understand the amount and magnitude I am projecting upon the reader's impression therewith to help her/him discern the very first motivations that drove my passion in this area to the current state which I find myself in and while writing this book and making the effort of documenting such a journey and also presenting it to the public.

In spite of the fact that the documentary itself has succeeded in leaving its mark on me, I yet wasn't able to extend my understanding – of the remarkable work these two gentlemen have presented – to the scope they themselves were examining in their own studies; let alone to be able to identify who Mr. Bauval was at the time and what role he played taking into consideration that Graham 'dominated' almost all the interviews and sceneries of that documentary and that I have programmed myself to set out – as I stated earlier – with the aim of finding a Journalist, not an Egyptologist.

I grew up not having or showing any interest in the Great Pyramid, or as a matter of fact, in any other ancient Egyptian *archeological* site. I remember looking at these monuments in the TV when I was a kid for the very first times and being aware of their existence; I also remember saying to myself: "*If I keep hearing narratives of graves rather than some Engineering mechanism that enables such structures to do something useful, I will never ever waste any more seconds looking at them!*". I couldn't articulate back then such words of course, even if I could, I certainly wouldn't be able to formulate them into such a sentence using the English language – which is not my mother tongue – nonetheless, this sentence describes exactly how I felt. The reader her-/himself can imagine what nonsense the *graves theory* presented to a young lad who was questioning everything before his eyes that did not deem convincing and consolidative with his own worldview; what exactly would a young boy think of people discussing about *mega structures* that exist in the world as being meaningless resting places for some Pharaohs? Well, what I myself was contemplating is that *either those people are idiots or that Pharaohs were some lame losers!* Later on I came to realize that I was wrong and these sites could have possibly served as tombs and also used for some other functions as well; it was '*Theology*' that gave me the turnaround route! As a growing Muslim person who was still young, I was sharpening my *Islamic Sword of Expression* against all of those who were daring and trying to take my faith away from me, so I was fully aware what Theology meant to man and which limit she/he is willing to reach by expressing her-/himself to protect and defend the metaphorical, vocal and linguistic grounds of belief as much as literally expressing the physical ones!

And We certainly sent into every nation a messenger, [saying], "Worship God and avoid satan". And among them were those whom God guided, and among them were those upon whom error was deservedly decreed. So navigate through the earth and observe how the end of the deniers was.
~ Quran 16:36

And that was how I started appreciating the message which such mega structures convey to us, but obviously not the mega structures themselves. This leaves the option for the *Mechanical Function* open for someone to capitalize on it and theorize its ancient existence; for that these monuments could have possibly served mankind in some *materialistic* way or another as well and, hence, they could be eventually looked at with appreciation. However, in this thesis, I am going to investigate one aspect of this message; the encoded part of it. The subject matter of interest in this *ciphered* information is surprisingly its mathematical part, that astounding portion is so magical that I am still being able to generate equations and relations thereof out of small amount of data, e.g., location and dimensions. I have taken the data from two major works in this field. The first one is from Robert's brother, Jean-Paul Bauval, and the second one is from Gary Osborn. But before I speak about the influence of Jean-Paul's and Gary's works on me, I will first introduce my contact with Graham Hancock and then with Robert Bauval to the reader.

After the exposure I've received – through the documentary – to the research of Graham Hancock and Robert Bauval, I started advancing my own theories and testing them myself through rigorous scrutiny and repeated examination using various data and information in the different fields of research that were related to that which I was tackling. The topic I was handling was based on my own observations; therefore, it was relatively new. And I say 'relatively' because its foundation has already been laid down by a preceding research which was published by another author – whose work I have completely elaborated upon in my book *32.5 System*. [2] But as you can tell, the spark of motivation that ignited all that enthusiasm started off only after I watched that documentary; and the culmination of that research ended up with Graham – after I have contacted him insistingly – telling me [3]

I've just read your article. You express yourself well but the subject is not for me .. Ultimately, I was not persuaded by your argument.

I wasn't discouraged by that; but to the contrary, it gave me a lot of inspiration to persistently assert that what I have had to say and also bring it forth to the public domain to harvest scrutiny as much as I could. I did not mention to the reader yet that I list Graham Hancock among the public figures that have influence on me; therefore, the reader should know by now that I do take Graham's feed in the most constructive way possible and integrate it with my own views. This has, after all, allowed me to embark upon such an opportunity to enter this field of research, i.e., the alternative.

Three days later from the date on which I received Graham's response, I published my very first book on Amazon [4] including the entire article which I presented to Graham along with some other conclusions I have reached. Since then, I had set myself on the course of publishing my research, editing and reviewing it repeatedly until I had completely documented my complete endeavors and thoughts in *32.5 System*. This is at least what I thought I have been doing by disclosing all of my work on this field, but I was yet again wrong! The research did not stop there and I was able to generate and extract more information therein/therefrom. Then I realized that I must release a new book edition which I am working on nowadays in addition to writing this book.

During that time of building/theorizing a platform under the discoveries and observations I have made, composing my thoughts onto paper, contacting Graham, publishing, editing, reviewing and scrutinizing, I "directly" felt the need for public exposure! I started adding people on facebook to my friends' list who were posting about relevant topics and those who were active on Graham's facebook page hoping somehow to engage them with my work. What made me quite frustrated in the beginning is that I couldn't find a facebook page for Robert. All what he had, was his own account onto which I have neither access to his posts nor any privilege to add him to my own friends' list yet. Long before I started seeking more contact with interested people, professionals and experts on the *alternative path*, I was watching lots of videos for Robert and reading some few articles on his work. What attracted me the most to his contributions was his sincere quest for deciphering the ancient Egyptian mysteries and narratives. He represented to me the perfect researcher I can trust with the topic of Egyptology; I just wasn't interested yet in delving into that field and rather enjoying the disclosures from Robert's own work.

I have bought a couple of books for Robert before and therefrom I knew the guy is no less of a writer than his friend, Graham! As a matter of fact, they both coauthored some books together and one of those titles fell into my hands; that's how I started appreciating the written narrative coming from these two guys. But now I am on facebook, and my friends' list has grown to include lots of interesting people. People I have never heard about before, like Gary Osborn; Jean-Paul Bauval, the brother of Robert; and others who are either researchers or part of that alternative Egyptology community. And somewhere there in the *cyber* mess of communication and by the merit of having mutual friends, I was given access to Robert's account to add him as a friend; and yes, he accepted!

It took about a year before the very first spark got flared before my eyes and drew my attention to one of Robert's interactions on Facebook for being especially relevant and resonating with my work. Robert was embracing a contribution from Gary Osborn – which he presented in one of his slides – for its significance in interacting with one of the mysteries of the Great Pyramid of Giza. It included a calculation which had the *Speed of Light* in it. At that moment, I knew I was going to be part of this discourse, but I didn't know yet what role I am going to play in it, nor was I able to tell if I was going to discover new horizons in that field of research afterwards, as I certainly did. It was indeed the Speed of Light that captivated my mind and not the huge structure itself despite the fact that its base' dimensions were the cause for delivering that beguiling number. One or two weeks later, Jean-Paul published an article [5] on the King's Chamber (KC) which completely sucked me into Egyptology and specifically into the topic of the Great Pyramid of Giza.

Jean-Paul wrote his thoughts using simple sentences and paragraphs; he documented his work on KC in the GPG and allowed the numbers to speak for themselves; he identified the role that the numbers play in recounting their own narrative. In other words, I identified Jean-Paul as one of my own gang! And since then, this exposure has had quite the influence on my discoveries that I am about to start presenting to the reader in the next chapter. Although I was never intending on researching the GPG myself, let alone to directly link it to my own work, Jean-Paul's contribution has glued me to it ever since. The amount of information I was able to generate from the data in Gary's slides and Jean-Paul's article and GPG sketch [6], guaranteed fixating my attention and research on the GPG alongside the work I have been doing so far on the alternative path. Gary's work showed me that the theme of my title *32.5 system* is linked with the GPG; and Jean-Paul's work opened my eyes on the Great Pyramid of Giza itself.

CHAPTER REFERENCES

[1] http://www.imdb.com/title/tt0252165/
[2] 32.5 System: The Complete Series Fused, by Ibrahim Ibrahim – 04 Oct 2015.
[3] Private Correspondence – 6 April 2015.
[4] 32.5: The Divine Mathematics of the Calendar, by Ibrahim Ibrahim, (Book Series: 32.5 System) (Volume 1) – April 9, 2015.
[5]
Facebook Public Link:
https://www.facebook.com/robbo50333/posts/101536398114190 16
Academia Link:
https://www.academia.edu/24594091/THE_LOCATION_OF_T HE_KINGS_CHAMBER_IN_THE_GREAT_PYRAMID_versi on_2

[6] Public:
https://www.facebook.com/photo.php?fbid=1015362524656901
6

2 THE RELEVANCE OF KING'S CHAMBER

I have published[1] an article online [1] after reading Jean-Paul's paper on his brother's Academia profile, i.e. Robert Bauval.[2] In this chapter, you will be reading a newer version thereof with extra commentary and elaboration. Its body extends here from the title below to the end of the chapter. The announcement of a discovery which I am making in it is very authentic and is not to be found anywhere else.[3] This article was the first phase in my *annexation* process onto the 'meter' numerical relations of the GPG; more insights and citations thereupon will be presented in the next chapters.

---

[1] On April 2016.

[2] See Chapter 1.

[3] Since nobody else laid that claim before I did here. My assurance is not only based on my own scanning and surveying and on that of the other experts and professionals in the field whom I am in contact with and who witnessed my own declaration and proclamation thereof, but also on the fact that I am relating to the dimensions of the GPG and the KC therein using the measurement unit of *meter*; which is not a conventional thing to observe among Egyptologists or researchers in this field at all (*see Appendix I*).

DIVINE PROPORTIONS IN KING'S CHAMBER

ABSTRACT: The significance of King's Chamber of the Great Pyramid of Giza is not only demonstrated through its positioning inside the pyramid structure, but also through its size. The major contribution of this study is discovering that the *volume* of KC equals to

$$(100 \times \pi)$$

and the projection therefrom is in association with the pyramid's base area rather than its base perimeter.

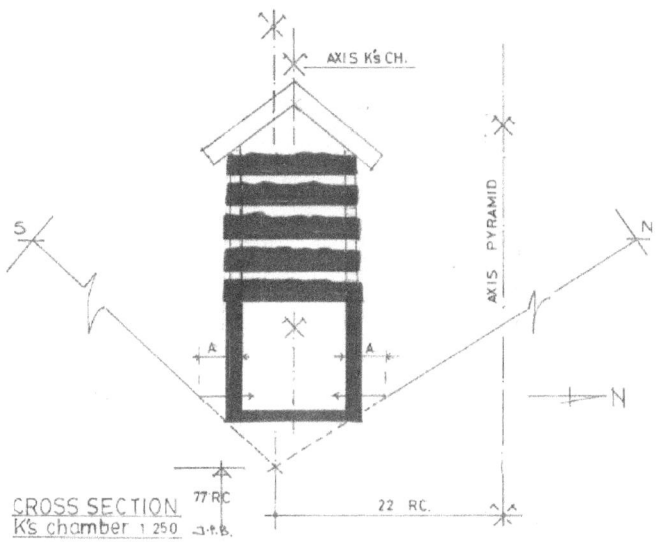

Figure 1

While viewing Robert Bauval's facebook post [2], I was directed through a link to an article which Robert's brother, Jean-Paul Bauval, has published on Academia [3]. Firstly, I noticed that the number '10.99' is not used consistently in both of the text and the illustration provided at the end of the article; a wrong value – probably a typo- which equaled to '11.99' showed up repeatedly in the body of the article. So I notified Robert who in turn updated his Academia feed accordingly and that is the reason why you see the text "(version 2)" in reference [3]. Secondly, I was drawn afterward into the narrative insomuch that I picked up my calculator and figured out therewith the volume of the King's Chamber based on the dimensions which Jean-Paul has provided. The value equals to 100 multiples of π with a precision of 99.78%.

$$\text{The Volume Of King's Chamber}$$
$$= 10.47 \text{ m x } 5.23 \text{ m x } 5.75 \text{ m}$$
$$= 314.859075 \text{ cubic metres} \dots (1)$$

Another significant proportion to take into account is that of the Great Pyramid's *apex angle* which is about

$$10.99 \times 7 = \left(77 - \frac{7}{100}\right) = 76.93° \cong 76.84°$$
$$= (180 - 2 \times 51.58°) \dots (2)$$

Where 10.99 is in meters and was recorded as such by Jean-Paul himself; it represents the distance between the pyramid's central axis and the KC. And

$$\frac{(440\,RC)^2}{2\left(280 \times \underbrace{\frac{1,100}{10}}_{area\ to\ perimeter\ ratio}\right)} = \frac{(440\,RC)^2}{2\left(280 \times \frac{77 \times 100}{7 \times 10}\right)}$$

$$= \frac{(440\,RC)^2}{100 \times 7 \times 2\left(\underbrace{28 \times \frac{11 \times 10}{7 \times 10}}_{=44}\right)}$$

$$= \frac{(44)^2}{7 \times 2 \times 28 \times (0.57 + 1)}$$

$$= \frac{22}{7} \times \frac{44}{28 \times (0.57 + 1)} = \pi$$

This also equals to

$$\pi = \frac{(44\ RC)^2 \times 10^2}{2(2.8 \times 1.1) \times 10^4} \rightarrow \frac{(44\ RC)^2}{2(2.8 \times 1.1)} = \pi \times 100 \quad \text{... (3)}$$

We also have from reference [4] (*on page 51*) the following relation for the whole Pyramid's structure

$$Length + Width - Height = 100 \times \pi$$
$$230.36 + 230.36 - 146.65 = 100 \times \pi \ \text{... (4)}$$

And if we replace 77 RC with the exact deduced value 76.93, we get the following:

$$\frac{76.93}{7 \times 10} \times 0.5236 = 0.57 = \frac{\pi}{2} - 1 \rightarrow \frac{76.93}{7 \times 10} \times 0.5236 \times 2.8 \cong$$
$$\varphi \equiv 1.618\ (in\ meters) \quad \text{... (5)}$$

The implication of using number '7' besides being a factor in the area-to-perimeter ratio, is that it leads to the following relation

$$\frac{440\ RC = 230.384\ m}{7} \cong 33\ meters\ \text{... (6)}$$

The number 33 contributes to the area itself as you will be reading below, but its significance can also be observed on the pyramid's height (Figure 2) as well. We also note the following proportion

$$\frac{103.514}{\pi} = 33 \cong \frac{230.384}{7} \rightarrow \pi \times 7 = 22 \quad \text{... (7)}$$

Equation (7) above highlights the significance of the *sectioning* by '7' (Figure 3) on the pyramid's base length. It further affects the base area by delivering 49 units (i.e., squares) of 7^2 square meters each; as a result of which, we get

$$49 \times 33 = 1618 - 1 = 1.618 \times 1,000 - 1 \quad \text{... (8)}$$

Seven multiples of these units (Area 1) deliver the figure of

$$1.618 \times 1,000 - 1$$

11

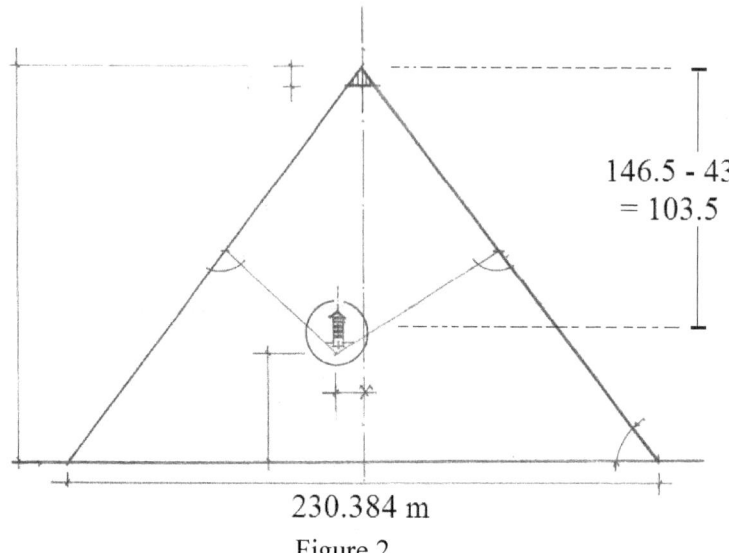

230.384 m

Figure 2

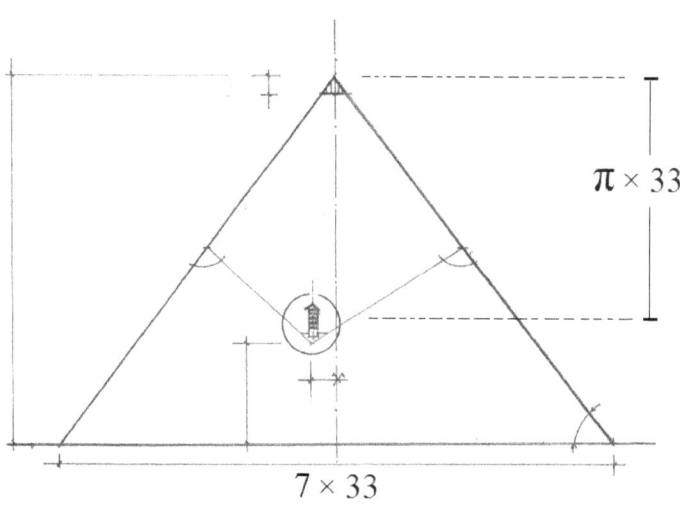

7 × 33

Figure 3

1	2	3	4	5	...	29	30	31	32	33
2					...					
3					...					
4					...					
5					...					
6					...					
7					...					

Area 1

And another 33 multiples thereof, will give the total area of the base. We saw above what role this 1,000 plays

$$\rightarrow \frac{7{,}700}{7} \times 0.5236$$

$$= [\, 1 + (10.99 \times 100) \,] \times 0.5236 =$$

$$= \left(\frac{\pi}{2} - 1 \right) \times 1{,}000 + 6$$

$$\rightarrow \frac{7{,}700}{7} \times 0.5236 - 6$$

$$= \left(\frac{\pi}{2} - 1 \right) \times 1{,}000$$

$$\rightarrow \frac{7{,}700 - 80.2139}{7} \times 0.5236$$

$$= \left(\frac{\pi}{2} - 1\right) \times 1{,}000$$

$$\rightarrow \frac{7619.7861}{7} \times 0.5236$$

$$= \left(\frac{\pi}{2} - 1\right) \times 1{,}000$$

$$\rightarrow \frac{(7 + 0.618 + 0.001)}{7}$$

$$\frac{\times 0.5236 \times 1{,}000}{7} + \frac{1}{17}$$

$$= \left(\frac{\pi}{2} - 1\right) \times 1{,}000$$

$$\rightarrow \frac{(7{,}000 + 618 + 1) \times 0.5236}{7} + \frac{1}{17}$$

$$= \left(\frac{\pi}{2} - 1\right) \times 1{,}000 \quad \dots (9)$$

The added value of '6' interestingly gets to be relevant only when I introduce a magnification measure of 1,000 (*you will also see it again in another equation shortly*), otherwise it is neglected because of being, in this case, just an unnecessary precision measure; nevertheless, the interpretation thereof is yet to be embarked upon in next chapters. However, we do see how it got distributed throughout the equation and another fraction came out as a result of that, i.e., '1/17' (*Its relevance will be presented in the next chapter*).

It is very exciting to also discern that the ideal model of a square delivers a unit diagonal of the Pyramid's base which equals to

$$\frac{1}{\sqrt{2 \times 440^2}}$$

$$= \frac{1}{7 \times \tan^{-1} 51.84}$$

$$= (1.618 \times 1,000 - 11) \times 10^{-6}$$

$$= \left(1.618 \times 1,000 - \frac{14}{\tan(51.84°)} \right) \times 10^{-6}$$

And from reference [5], we get

$$= \left(\frac{1.618}{1.57} \times 1,000 - 7\right) \times 1.57 \times 10^{-6}$$

$$= \left(\frac{1.618}{1.57} \times 1{,}000 - 7\right) \times 0.5236$$
$$\times \frac{Speed\ of\ Light}{100} \times 10^{-12}$$

$$= (103.057 \times 10 - 7) \times 0.5236$$
$$\times 2.99792458 \times 10^{-6}$$

$$\cong ((33 \times \pi - 1.618^{-1}) \times 10 - 7)$$
$$\times 0.5236 \times 2.99792458$$
$$\times 10^{-6}$$
$$\cong \left(\frac{(33 \times \pi - 1.618^{-1})}{10} - \frac{7}{100}\right)$$
$$\times 0.5236 \times 299.792458$$
$$\times 10^{-6}$$

$$\cong \left(\frac{(3 \times 110 \times \pi - 6.18)}{100} - \frac{7}{100}\right)$$
$$\times 0.5236 \times 299.792458$$
$$\times 10^{-6}$$

$$\cong a\ millionth\ of\ \sqrt{GPG's\ Volume}$$

... (10)

We can see here how the encoding of the Speed of Light value itself along with the

$$33 \times \pi$$

distance tune together the GPG's proportions to this relation while taking into account the conversion between the units of Royal Cubits and Meters. The *Preferred Choice of Measurement* using the distance of ($33 \times \pi$) is not random, and its significance will be elaborated upon in the next chapter.

Right rectangular pyramid
Solve for volume ▾

$$V \approx 2.59 \times 10^6$$

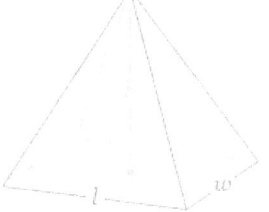

l Base length   230.384

w Base width   230.384

h Pyramid height   146.608

Solution

$$V = \frac{l\,w\,h}{3} = \frac{230.38 \cdot 230.38 \cdot 146.61}{3} \approx 2.59383 \times 10^6$$

This amazing relation also delivers, as you can see above, a difference in value between two numbers that resemble the circumferences of the two circles of the Pyramid's base. Since this equation serves to iterate the diagonal one unit of length in multiples of the Speed of Light value, the difference between the two circles is maintained at this constant value. The significance of such a design lies in the fact that it is the location of King's Chamber along with the introduced sectioning of the Pyramid's apex which uncloak together, based on the Royal Cubit, this nature of the Great Pyramid's geometry.

Note that the figure of 6.18 minutes of arc equal to 103°/1,000 and from reference [6], we see that the King's Chamber lies in-between the two shafts that meet down in that angle of 103°. I had the luxury of having Robert interact with my posts in this regard with the following statement:

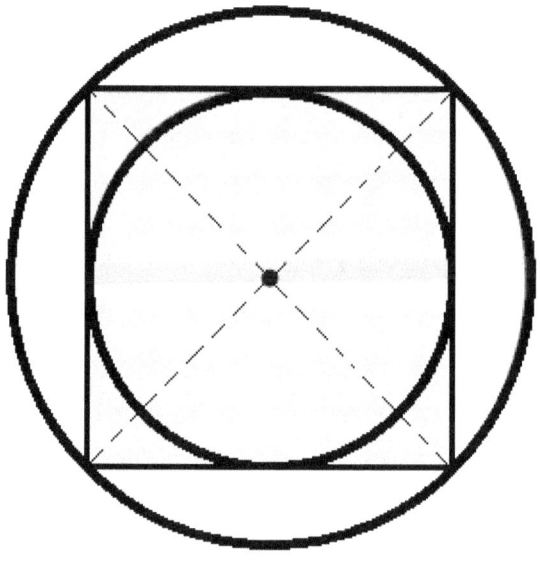

Figure 4

"Gantenbrink gives the slopes in of KC shafts as 45 and 32.47 (in degrees). This comes to 77.47, making the angle between the shafts 180-77.47 = 102.53. Petrie, however, gives 45.21 and 31.55 = 76.76, making the angle between them 180 - 76.76 = 103.24. Now according to Gantenbrink the shafts emerge at a height of 154 RC which, according to Petrie, is just above the 103rd level of the pyramid."
~ Robert Bauval [7][8]

Therefore, not only the vertical location of this Chamber is relevant, but the two shafts' angle that encloses it was engineered to **conserve the expression** of the Speed of Light value in the Pyramid's base diagonal. I am coining this phrase, i.e., *conserve the expression*, intentionally to point out the significance of such a design insomuch that even small details do matter until they get also fully elaborated upon and in turn.

Another thing to discern in those rich equations is the figure of 51.84 showing up as a measure of an angle as well as a measure of a distance:

$$7 \times \tan^{-1} 51.84$$

in the latter case, we see alongside it a magnification amount of 7 which accompanies it to adjust the design of the pyramid's base dimensions. This magnification value enlarges the ratio of the opposite over the adjacent sides by seven times.

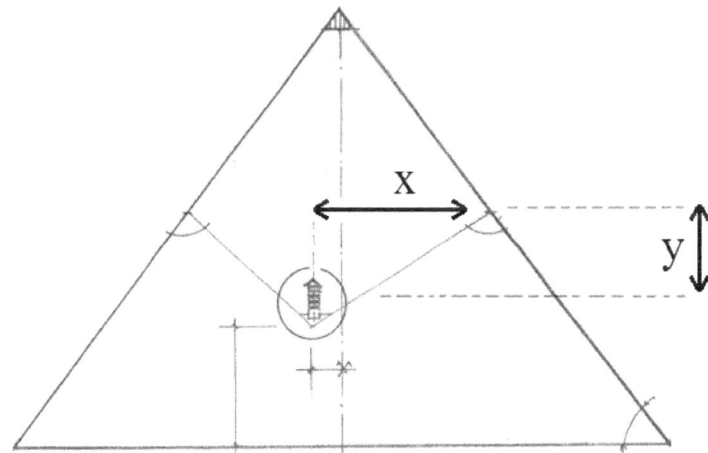

Figure 5: KC is circled to mark the whole location but y actually extends down to the shafts' conjunction.

From Jean-Paul's sketch [9], we know that x = (22 RC + half the distance between shafts' openings); and from my own research (*you will read about it in the next chapter*), we have (half the distance between shafts' openings = $33\pi/2$).

Therefore,

$$x = 100/1.578 \approx 100 \times 2/\pi \ldots$$
$$(11)$$

And we also saw earlier the significance of the 100 in this construct:

$$\left(\frac{(3 \times 110 \times \pi - 6.18)}{100} - \frac{7}{100} \right)$$

Surprisingly, this magnification figure of 7 produces a difference of

$$32 \times 1{,}000 \; minutes \; of \; arc$$

between the fundamental angle

$$\tan^{-1} 51.84$$

and the newly generated value thereof, i.e.,

$$7 \times \tan^{-1} 51.84$$

The angular preservation of the dimensions even extends to the RC units of the base diagonal length

$$\frac{1}{\sqrt{2 \times 440^2}} = \frac{1}{622}$$

Where

$$\frac{360}{622} \approx \frac{\pi}{2} - 1 \approx \frac{1}{1.732} \ldots (12)$$

CONCLUSION

It is now crucial to take into account that the dimensions of the Great Pyramid of Giza are in the right proportion with the size of the King's Chamber. Therefore:

A. Not only the position of KC is significant, but its size also matters.
B. It is not the perimeter of the base which is important, but rather its area.

ACKNOWLEDGEMENTS

I sincerely thank Robert Bauval for all his contribution so far in the different fields of studies in Egyptology and especially for caring to share his work with the public; this is how I got to enter this specific field of research after all. I also wish to give special thanks to his brother Jean-Paul Bauval whose article influenced me enough to write this paper and enabled the major discovery of KC's significant size. I wish indeed to embark upon the opportunity of coauthoring an article with him in future. And I would also like to show my gratitude to Gary Osborn whose Facebook Slides (April 2016) [10] gave me the very first access onto the fact that the Pyramid's base circles have a difference in value which is proportional to the Speed of Light value; upon such an insight I was able to devise my way in calculating the one unit length of the diagonal and also to account for the missing 10^6 factor in his slides and in Scott Onstott's book (see reference [4]). I took the reference to the shafts' angle of King's Chamber, i.e., 103°, from Gary's slides as well. [11]

SOURCES REFERENCED IN ARTICLE

[1] https://www.academia.edu
/24605088/Divine_Proportions_in_Kings_Chamber
[2] *Public*: https://www.facebook.com
/robbo50333/posts/10153639811419016
[3] https://www.academia.edu
/24594091/THE_LOCATION_OF_THE_KINGS_CHAMBER_I

N_THE_GREAT_PYRAMID_version_2

[4] *Quantification: Illustrations from the Creator of Secrets In Plain Sight*, Scott Onstott, 04 April 2014.

[5] https://www.academia.edu /24852786/Royal_Cubit_and_Light_s_Speed

[6] *Public*: https://www.facebook.com /photo.php?fbid=10208074894924717

[7] *Private*: https://www.facebook.com /32point5/posts/1598278973819108?comment_id=15984606204 67610

[8] *Private*: https://www.facebook.com /32point5/posts/1598278973819108?comment_id=15984649071 33848

[9] See reference [3]

[10] *Private*: https://www.facebook.com /photo.php?fbid=10207890777001884

[11] See reference [6]

FIGURES INCLUDED IN ARTICLE

1) The Location Of The King's Chamber In The Great Pyramid (version 2), by Jean-Paul Bauval.
   Link: see reference [2]
2) Ibid. (*I've modified its text*)
3) Ibid. (*I've modified its text*)
4) (*my own sketch*)
5) See (1). (*I've modified its text*)

TIMELINE OF DISCOVERIES IN ARTICLE

First article edition was published on the 20th of April 2016 along with the discovery of the volume being 100π and the relevancy to pyramid's base area instead of perimeter. Then few days after that, the figure of 33π was located in the GPG along with the Speed of Light calculations that mark the base diagonal. And on 22 May I discovered the link to the pyramid's volume. The final text addition into the article was made on the 24th of May culminating in the discovery of 51.84 sides' ratio.

3 TIMELINE OF DISCOVERIES

Herein I list my discoveries which I had pronounced on my facebook account chronologically:

- **22 April 2016**

> The value equals to
> 100 multiples of π with a precision of 99.78%.
>
> > The Volume Of King's Chamber =
> >
> > 10.47 m x 5.23 m x 5.75 m = 314.859075 cubic metres

As I have pointed out earlier, the volume of the most important part in the Great Pyramid of Giza is significant and relevant to the notion of Divine Proportions that are incorporated in this structure.

- **27 April 2016**

$$\frac{2.99792458}{1.57} = 1.9095 = \frac{1}{0.5236}$$

The ancient egyptian Royal Cubit -with reference to the metric system- equals to the ratio of Light's Speed (in 10^8th of a second) over 1.57. The latter number is just the ratio of Great Pyramid's base width to its height.

Gary Osborn interacted with this post and pointed out that $\pi/6 = 0.5236$

(More data and information on the same topic is in Appendix III)

- **29 April 2016**

I have discovered earlier the distance of 33π in the GPG's cross sectional structure relative to the KC's location as mentioned in the article, but on this date I was able to locate it in the pyramid's base diagonal length as well.

$$\left(\frac{(33 \times \pi - 1.618^{-1})}{10} - \frac{7}{100} \right) \times 0.5236 \times 299.792458$$

Outer Circle   Inner Circle

The equation above delivers one single unit of length from the GPG's ideal base diagonal; hence, the Speed of Light value iterates step by step accordingly to produce the desired length of the diagonal while maintaining the difference between the two circles as a constant value. The vertical position of KC and the introduced angle-sectioning in the GPG's apex uncloak together based on the Royal Cubits tis nature of the GPG's geometry.

Gary Osborn interacted with this post and pointed out that
A.
    280 Royal Cubits / 2.718 (e) = 103.01692420898
B.
    33 π = GPG's side angle × 2
C.
    The Two Shafts Angle = 33π = 103°

So, I annexed the measure of distance in meters and Gary reacted thereupon by discovering it as an angle; not just any random angle, but rather the most important one in the structure: The one that encloses the King's Chamber. This is the first time when Gary modified his slides and added my contribution thereon giving me credits thereby; thanks Gary!`

- **30 April 2016**

After Gary's contribution in relating my 33π figure with the units of degrees, I found out that the magnification factor of 1,000 – which I have referred to in the article – alongside Gary's units of degrees also exist in the pyramid's base dimensions as well.

One unit length of Great Pyramid's base diagonal equals to

$$\left(\frac{(3 \times 110 \times \pi (- 6.18))}{100} - \frac{7}{100} \right) \times 0.5236 \times 299.792458$$

Where 6.18 minutes of arc equal to $103°/1{,}000$

And $\quad \dfrac{(3 \times 110 \times \pi - 6.18)}{100} = \dfrac{103}{10}$

This signifies the relevance of
the 1,000 magnification factor

Thereafter, I also discovered the pyramid's base dimension with the celestial mechanics relevant to man's observation and calendar system:

The angle travelled by the Earth-Moon system around the Sun during one sidereal month of period 27.321661 days

$$\theta = (27.321661 \times 360°) / 365.25636 = 26.92848$$

While one unit of diagonal length of the Great Pyramid's base equals to

$$\frac{1}{\sqrt{2 \times 440^2}} \quad RCs \;=\; \frac{26.78°}{10^6}$$

365.25636 is a tropical year with the reference to the stars on the ecliptic (i.e., the apparent path of the Sun). Note that 103 / 26.78 = 100 / 26 and also that it takes 100 sidereal moon cycles (i.e., 27.32) to cover that whole one million factor based on a 366 System (more on this in my book *32.5 System*). It is known though that the length of a tropical year is the time it takes the Earth to complete a full orbit around the Sun, but it varies from year to year; therefore there exists no exact number for the tropical year period.

- **01 May 2016**

$$\frac{27.32}{29.53} \times 360 = 333 + (\frac{\pi}{2} - 1)/10$$
$$= 1{,}000/3 - 27.32/100$$
<small>(99°₀ precision)</small>

The 1,000 magnification factor comes from the ratio of the sidereal over the synodic months for a period of a year's quarter. In other words, it is made by the following angular drift

$$= 360 - (26.6 \sim 26.9)$$

The 366 System denotes the delay within the year's quarter period because of the lunar different periods, and that delay equals to the Megalithic Yard. And from the previous chapter we relate to 17 by the following

$$\frac{27.32}{29.53} \times 360 = 333 + \frac{100}{1{,}700 + 26.9}$$

$$= 333 + 0.057$$

Where

$$\frac{1}{17} = 0.0588$$

And

$$0.57 \times 57 \approx 32.5$$

$$0.57 \times 57.88 \approx 33$$
$$0.57 \times 58 \approx 33$$
$$0.57 \times 58.8 \approx 33.5$$

- **05 May 2016**

Robert Bauval told me:

> *I do appreciate your work.*

This is when **I** commented back:

> *When recognition comes upon you from an Authority,*
> *celebration offers you a sip of accomplishment and tranquility.*
> *Thank you Robert Bauval!*

- **06 May 2016**

$$\frac{1}{\sqrt{2 \times 440^2}} \; RCs \;=\; \frac{26.78°}{10^6}$$

where

$$1{,}000{,}000 \; \text{Mod} \; 360 = 280$$

The excess of angular rotation equals to the Great Pyramid's height.

$$1{,}000{,}000 \; \text{Mod} \; 366 = 88$$

88 is the officially recognized number of constellations

Notice that

$$\frac{88}{280} = \frac{\pi}{10}$$

And

$$1{,}000 \; mod \; 360 - 1{,}000 \; mod \; 366 = 12$$

$$\frac{1{,}000{,}000}{360} = 2777.7777$$

$$\frac{2777.7777}{26.78°} = 1/103°$$

This is when Gary Osborn discovered that

10 x pi in meters = 31.4159265358979324
= 103.07062511777536 feet

And also presented to me that

$$5 \times 0.5236 \; meters = 2.618 \; meters$$

The number 2.618 is important since

$$1.618^2 = 2.618$$

- **08 May 2016**

After re-viewing Jean-Paul's sketch, I discovered that the cross-sectional distance that separates the two shafts' openings (121+77=198 RC) equals to 103.67 meters. I thank Gary Osborn yet again for modifying his slides for the second time while giving me credits on this other discovery as well. The significance of my observation is noticing the relevance of the units of meters in the structure of the Great Pyramid of Giza.

Robert Bauval interacted with this post saying:

The angle of the KC north shaft is not 32 deg. but 32.47 deg. This is according to Gantenbrink's assertion that the tangent of the KC shaft is 7/11 = 0.636363 which is 32.47 deg. and thus making the obtuse angle between the two shafts to be 102.53 deg.

I responded:

32.47 deg. for the axis of the shaft? We still do have a margin of movement though depending on the shaft's area sides, right?

Robert:

Actually, if you take Petrie mean values for the KC shafts, which are 45.21 + 31.55 = 76.76..... then 180 - 76.76 = 103.24.... which is closer to the 121 RC and 103.67 m. ☺

I:

so we are good Sir? ☺ ☺

Robert:

Yes, I think so.

But then **Robert** commented giving significance to this discovery by saying:

This may be used as another good argument that shows a

relationship between the RC and the meter, as well as the 'counting' of the course levels.

Robert:

And Ibrahim, please stop calling me 'Sir'. Robert will do just fine.

Jean-Paul:

same goes for me ... JP or Jean-Paul ... will do.

LINKING THE GROUND RANGE RESOLUTION TO π

The other discovery I have made on that day was when I noticed that if we to calculate the *Ground Range Resolution* of the Pyramid's shafts' openings (for one single pulse), we find that

$$33\pi \times 2\cos 32.47 = 100/(\frac{\pi}{2} - 1)$$

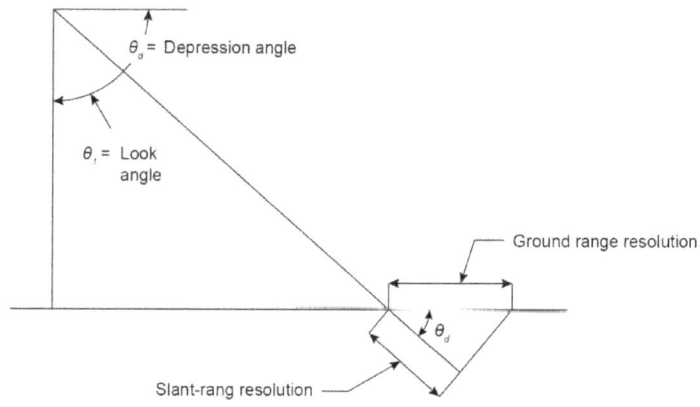

Figure 1: Ground Range Resolution

What this equation is saying is that ($\frac{\pi}{2} - 1$) which shows up in our calculations all the time is an *Image Resolution* Factor. And for the other shaft's angle (i.e., 45°), my book *32.5 System* addresses that number which is in reference to the precession itself; you will further read about it in Chapter 4.

Please note that I am not saying that the shafts themselves receive radar pulses in them from the same source simultaneously, but rather, they were designed accordingly as if they were to expect some sort of pulses!

- **16 June 2016**

After looking at the latest slides which were posted by Gary Osborn on his facebook wall, I noticed the relevance he gave and observed in the number 1.2732; and thereupon I discovered the four following relations:

THE RELEVANCE OF GARY'S NUMBER (1.2732) TO MY NUMBER (1.68)

$$\frac{1,000}{1.68} = 60 \times \pi^2 \rightarrow 6 = \frac{10,000}{168 \times \pi^2}$$

$$0.6 + \frac{168}{10,000} \cong \frac{10,000}{127.32^2} \cong \frac{1}{1.2732^2}$$

$$\frac{33}{32.5} \times \frac{100}{168} + \frac{168}{10,000} \cong \frac{1}{1.2732^2}$$

THE RELEVANCE OF NUMBER 7 TO THE ROYAL CUBITS

$$7 \times 0.5236 = \pi + 0.5236$$

THE LINK BETWEEN THE ANGULAR DRIFT OF THE CELESTIAL REFERENCE AND THE ROYAL CUBITS

$$26.9 \cong 360 \times \frac{0.5236}{7}$$
$$= \left(1 - \frac{27.32}{29.53}\right) \times 360$$

THE RELEVANCE OF THE 10,000 FACTOR TO THE NATURAL LOGARITHM AND π

$$1.273 = e^{\frac{1}{\pi+1}}$$

But most importantly

$$e^{\frac{1}{\pi+1}} = e^{\frac{100}{27.32} - \frac{100}{29.25}}$$

(See Appendix III)

MEASURING THE DISTANCE TO THE HORIZON

I was intrigued so much by Gary's number (1.2732) that I researched thoroughly to investigate the implication of its presence among all other numbers in the different by-us devised models of interpretation and succeeded in finding out its relevance to the horizon.

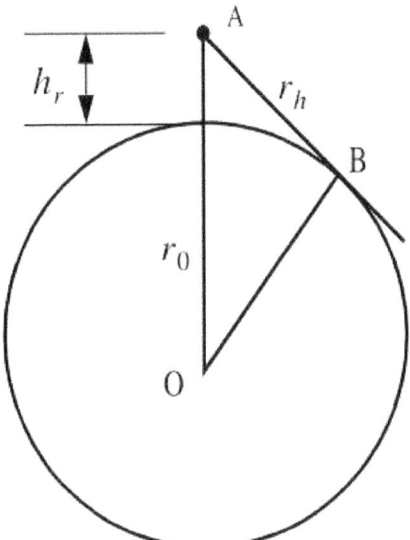

Figure 2: The Great Pyramid of Giza and the Horizon

$$r_h = \sqrt{(r_o + h_r)^2 - r_o^2}$$

$$approximately, r_h = \sqrt{2h_r r_o}$$

These are idealized conditions (by ignoring atmosphere refraction). And by substituting Earth's radius

$$r_o = 637,1 \ km$$

I arrived at

$$r_h = 27.32 \times 1{,}000$$

$$h_r = \pi \times \frac{186.4}{10}$$

These numbers are significant foremost for the following relation (which I have embarked upon)

$$27.32 \times \frac{1{,}000}{1.2732} = 146.4^2$$

where 146.4 corresponds to the pyramid's height, and also for agreeing with Gary's devised proportions in the following figure (which he presented on his facebook wall)

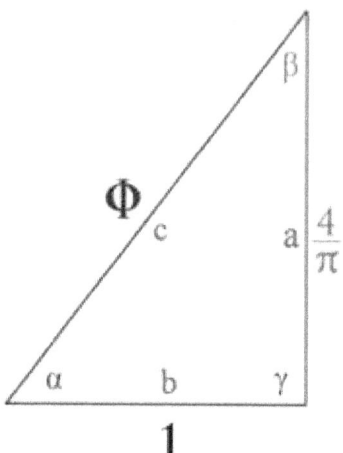

Figure 3: Phi Proportions [1]

where the length, c, equals to 1.618 meters after dividing 186.4 over the pyramid's base apothem.

Another interesting relation I observed was when substituting with

$$h_r = 280(0.5236)$$

Which delivered

$$r_h = (6 + \frac{\pi}{2} - 1)^2 \times 1,000$$

- **20 June 2016**

After I have discovered the relation between the GPG's engineered proportions with the horizon, I was also able to identify the following:

$$\frac{186.4 \times \pi}{10} = \frac{146.4}{2.5}$$

$$2.5 \times 146.4 = 366 = \frac{10,000}{27.32}$$

$$186.4 \times 2.5 = 366 + 100$$

$$= \frac{1.2732 \times 10,000}{27.32}$$

One can further read more about the relevance of '2.5' in *32.5 System*.

Gary has made a comment on my above posts by contributing to these relations with the following

$$\frac{186.4}{146.4} = 1.2732$$

- **24 June 2016**

THE CUTOFF FREQUENCY OF A WAVEGUIDE

Rectangular and cylindrical waveguides are only two of an infinite variety of forms in which single-conductor hollow waveguides can be made. The waveguide could have an elliptical cross-section or a reentrant one as shown below. [2]

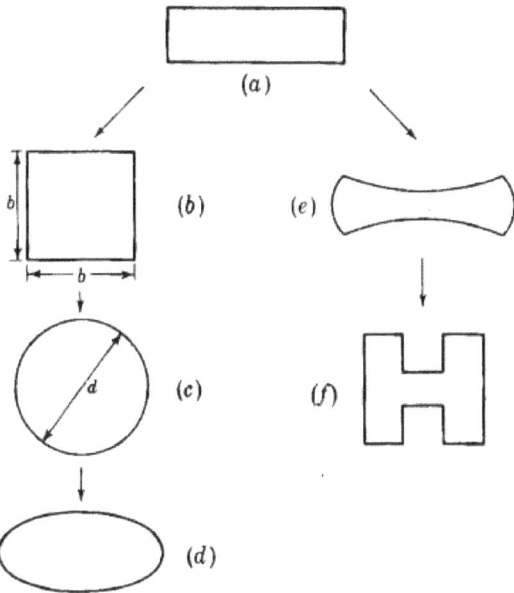

Forms of hollow single-conductor waveguides showing how square (*b*), cylindrical (*c*), and elliptical (*d*) shaped waveguides may be derived from rectangular type at (*a*). Types (*e*) and (*f*) may also be derived from (*a*).

Figure 4: Deriving Different Shapes of Hollow Waveguides from the Rectangular Waveguide

And what is interesting here is that the longest wavelength that the circular waveguide can transmit (aka, cutoff wavelength) is:

$$\lambda_c = \sqrt{\pi} \times d = 1.77d = 3.54r$$

Substituting the diameter with the significant cross-sectional length of the pyramid into the above formula delivers

$$\lambda_c = 1.77 \times 33 \times \pi$$
$$= (1 + 0.83) \times 100$$
$$= 366/2$$

Yet more precisely

$$\lambda_c = \sqrt{\pi} \times 33 \times \pi = \frac{365 + 2.5}{2}$$

Substituting, however, the radius with the same value delivers

$$\lambda_c = 3.54r = 3.54 \times 33 \times \pi$$
$$= 366 + 1$$

(See Appendix V for more Info on 1.83)
(More on Electromagnetics Engineering in next chapter)

CAVITY RESONATORS

The purpose of transmission lines and waveguides is to transmit electromagnetic energy efficiently from one point to another. A *resonator*, on the other hand, is an energy storage device. As such it is equivalent to a resonant circuit element. At low frequencies, a parallel-connected capacitor and inductor, as in the figure (a) below, form a resonant circuit.to make this combination resonate at shorter wavelengths, the inductance and capacitance can be reduced, as in (b). Parallel straps reduce the inductance still further, as in (c). The limiting case in the completely enclosed rectangular box, or cavity resonator, as in (d). In this cavity resonator the maximum voltage is developed between points 1 and 2 at the center of the top and bottom plates. [3]

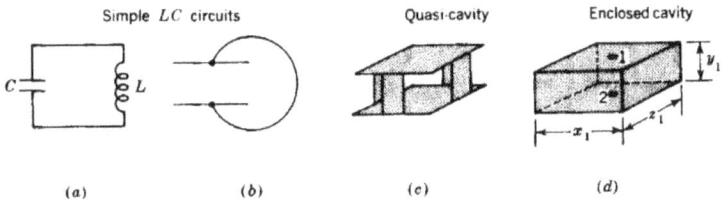

Simple LC circuits   Quasi-cavity   Enclosed cavity

(a)   (b)   (c)   (d)

Evolution of (enclosed) cavity resonator from simple LC circuit.

Figure 5: Evolution of The Enclosed Cavity Resonator

The basic principle of a cavity resonator can be described in connection with the pure standing wave. Here the energy oscillates back and forth from entirely electric to entirely magnetic twice per cycle. [4]

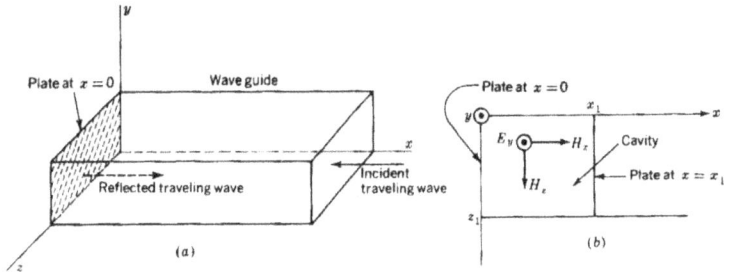

(a) Perspective view of rectangular waveguide closed with plate at $x = 0$ and (b) cross-sectional top view with additional plate at $x = x_1$ trapping wave inside the cavity.

Figure 6: A Enclosed Rectangular Box

With no wave propagation in the y direction as seen in the figure above, we have the relation

$$\lambda = \frac{2}{\sqrt{(l/x_1)^2 + (m/z_1)^2}}$$

where λ is the resonant wavelength, l refers to half-cycle variations of the fields in the x direction and m in the z direction. And when $x_1 = z_1$, [5] we get

45

$$\lambda = \frac{2}{\sqrt{(2/x_1)^2}} = \sqrt{2}\, x_1 = 1.41\, x_1$$

But note that:

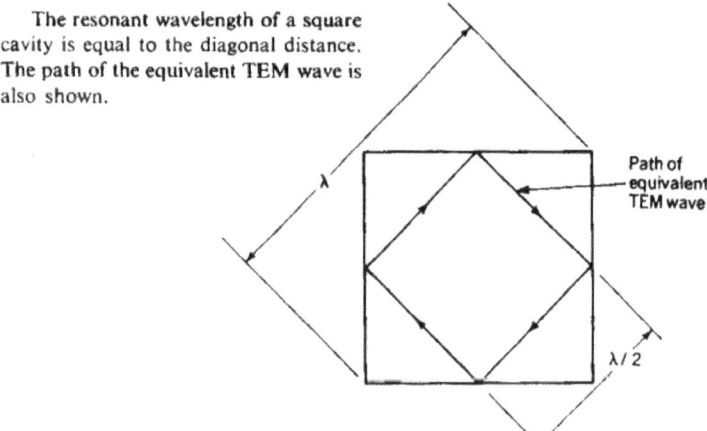

The resonant wavelength of a square cavity is equal to the diagonal distance. The path of the equivalent TEM wave is also shown.

Path of equivalent TEM wave

λ

λ / 2

Figure 7: The Resonant Wavelength of a Square Cavity

This means if we substitute the 33π into x_1, we will arrive at

$$\lambda = 33\,\pi \times 1.41 = 146.6$$

Which equals interestingly to the height of the Great Pyramid in meters! And if we substitute the width of its base, 440(0.5236), into x_1, we will arrive at

$$\lambda = \sqrt{2} \times 440(0.5236) = 325$$

(See next chapter for the relevance of 325)

However, what is really astounding in all this is when we consider the units of Royal Cubits in the above equations. In other words, the relation delivers an unexpected value, which is:

$$\lambda = \sqrt{2} \times 440 = \frac{1}{base\ diameter\ unit\ length}$$

This is amazing in every sense! So what exactly is the message being conveyed here? Let's inspect these relations further and see what we get when treating the dimensions as if they were that of an antenna.

- **31 August 2016**

THE SIGNIFICANCE OF THE ANGLES OF THE UPPER SHAFTS IN THE GREAT PYRAMID OF GIZA

We have seen (at the end of Chapter 2) how the major angle in the Great Pyramid is linked with the base diagonal length, and here we examine how it is linked with the angles of the two upper shafts through that same length.

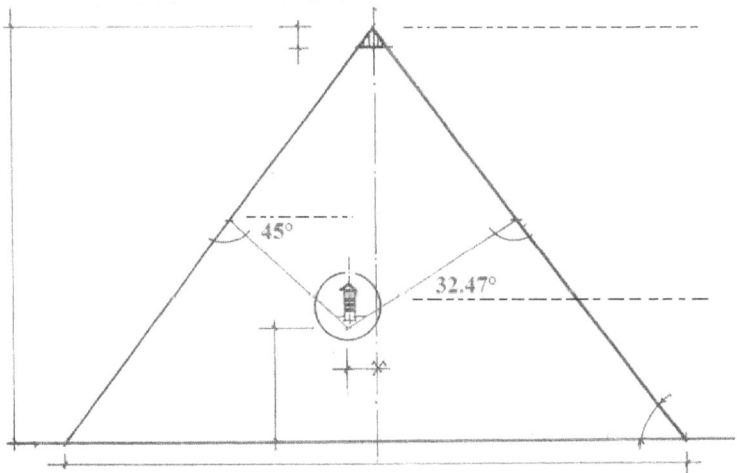

Figure 8: The Angles of the Two Upper Shafts in the Great Pyramid of Giza

Robert Bauval vouches for Gantenbrink's work who was the Engineer who has measured both of these values for the two upper shafts – as I have mentioned in Chapter 2. Robert has written on his website [6] the following praise for Rudolf Gantenbrink

" Rudolf Gantenbrink (is) perhaps the only man alive I know who truly has explored and analysed every detail of the shafts for many years now. "

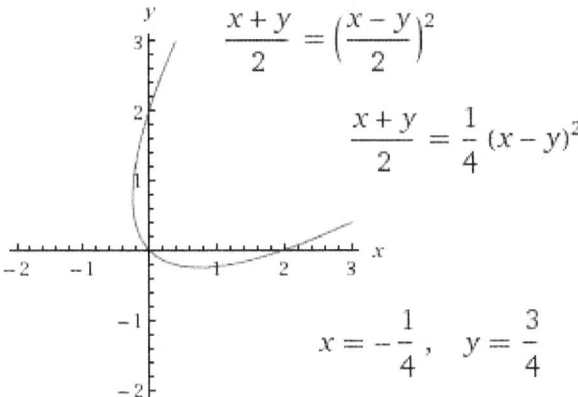

Figure 9: The Equation which Links the Angles Together

The angles of the upper two shafts in the Great Pyramid of Giza act together as one. They were chosen carefully to reflect the length of the base diagonal insomuch that they were tweaked to conserve the value of the difference between them in addition to the average thereof. I am sure that this was an ingenious solution which were devised to enable some Engineering application that uses both of the shafts in its process!

CHAPTER FIGURES

1) Remote Sensing and Image Interpretation, 5th Edition by Thomas Lillesand (page 647).
2) Radar Systems Analysis and Design Using Matlab, by Bassem Mahafza.
3) Gary Osborn.
4) Electromagnetics with Applications, 5th Edition by John D. Kraus (page 473).
5) *Ibid* (page 492).
6) *Ibid* (page 493).
7) *Ibid* (page 494).
8) The Location Of The King's Chamber In The Great Pyramid (version 2), by Jean-Paul Bauval (*I've modified its text*).
9) (*Produced using WolframAlpha*)

CHAPTER REFERENCES

[1] Gary Osborn, FB Slides – *Private:*
https://www.facebook.com/photo.php?fbid=1020835414470578
7
[2] Electromagnetics with Applications, 5th Edition by Kraus.
[3] Ibid.
[4] Ibid.
[5] Ibid.
[6]
http://www.robertbauval.co.uk/articles/articles/scsequel2.html

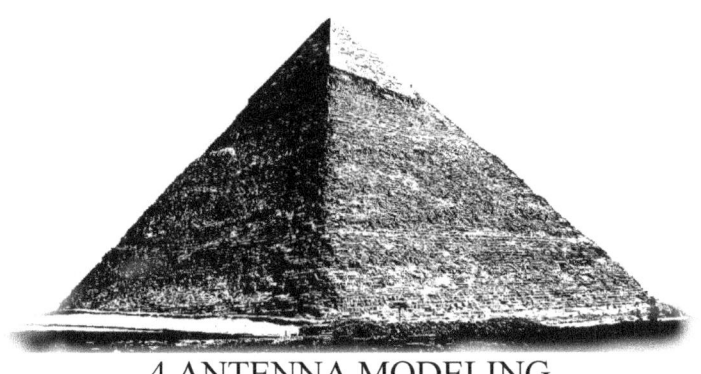

4 ANTENNA MODELING

T he directivity of circular loop antenna as a function of loop circumference in wavelengths is demonstrated in the below graph. Note that for short circular loop antennas, the graph starts as a constant value of 3/2, and then it increases gradually for longer circular loop antennas steadily by a factor of 0.68. *(See Appendix VI)*

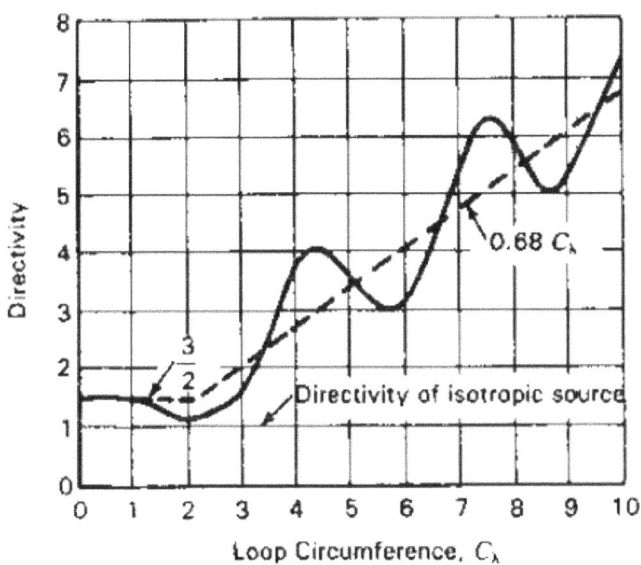

Figure 8: Antennas, 3rd Edition by John D. Kraus (page 254)

And in another reference we see that the Radiation resistance and directivity for circular loop Antennas is demonstrated in the following graphs

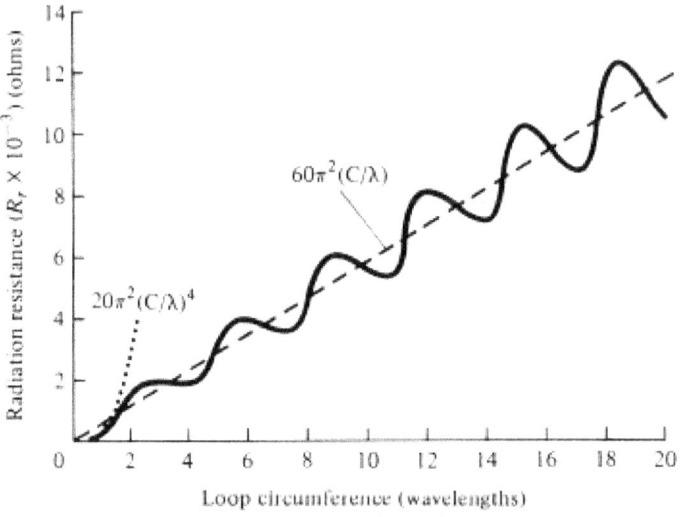

(a) Radiation resistance of circular loop

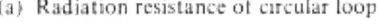

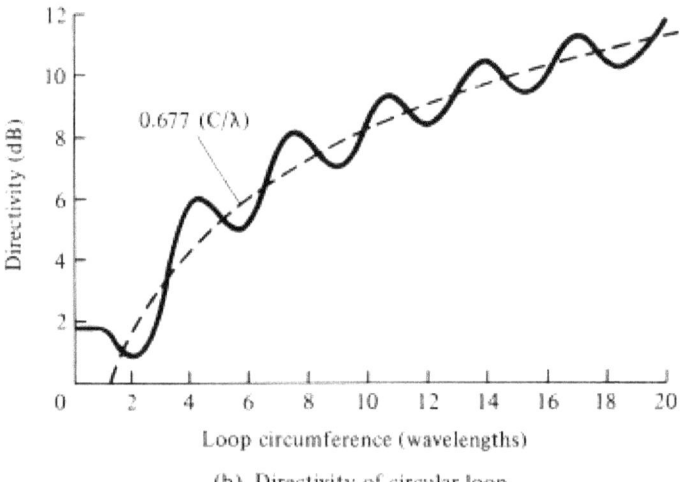

(b) Directivity of circular loop

Figure 9: Antenna Theory - Analysis and Design, 3rd Edition by Constantine A. Balanis (page 254)

What the figures above are showing is that the number 0.68 (in Fig. 1) or its rounded-off value of 0.677 (in Fig. 2) are so relevant in the design of Loop Antenna inasmuch as it is accompanying the length of the wavelength variable in determining the directivity of this specific kind of Antennas. This number is linked with the Image Resolution equation which I have presented in Chapter 3 where we have

$$33\pi \times 2\cos 32.47° = 100/(\frac{\pi}{2} - 1)$$

$$\rightarrow 2\cos 32.47° = \frac{100}{\frac{\pi}{2} - 1} \times \frac{1}{33\pi}$$

$$= 1.68 = 1 + 0.68$$

And for the other angle, we have

$$33\pi \times 2\cos 45° = 10,000/68$$

Where

$$\left(\frac{\pi}{2} \times 10\right)^2 = \frac{10,000}{68} \times 1.68$$

$$= \frac{1}{\sqrt{2 \times (33\pi)^2}} \times 1.68$$

The latter equation resembles an extension of small area units of

$$4 \cos 45° \cos 32.47°$$

Giving thereby the important relation of

$$\frac{\left(\frac{\pi}{2} \times 10\right)^2}{4 \cos 45° \cos 32.47°} = 33\pi$$

$$\approx (60 + 1.68) \times 1.68$$

$$\rightarrow \frac{33\pi}{1.68} \approx (61.8 - 0.1)$$

$$= \frac{10,000}{161.8} - 0.1$$

We notice here that even if the Image Resolution shrunk down to $2 \cos 32.47°$ instead of extending all over the distance of 33π, its value will remain preserved such that 60 minutes of arc almost equal to $100 \times (\frac{\pi}{2} - 1)$ radians^{-1}; this accounts for the 10,000 magnification value which is the result of the imaginative diagonal length of

$$\frac{1}{\sqrt{2 \times (33\pi)^2}}$$

But is this one unit length of a diagonal really an imaginative value? It is certainly NOT if we take into consideration the cross sectional area of that course level regardless of the absence of any shafts' openings on the other sides of that cross-sectional horizontal square. The reader will be amazed to know that such an application actually exists in today's Engineering systems, hence, the title of this Chapter. Despite that I am not concerned with any conductive material layers being applied on some surfaces of the pyramid to at least enable the structure to act as a real Antenna, this is nonetheless irrelevant since I am not asserting that the GPG operated as such a device, nor am I excluding the possibility thereof either. If the numbers are delivering some sort of a message that, in turn, narrates some "static" story of an application in Engineering, we certainly shouldn't just drop off the whole investigation as a mere coincidence. If we are still able to write down proper *vivid verbs* to link the different *exanimate nouns* together, then a narrative of some kind is indeed being sculptured and awaiting to be fully uncloaked one day.

ANTENNA DESIGN AND PYRAMID'S DIMENSIONS

Horn Antennas are famous for their simplicity and wide spread applications in the microwave spectrum. The most widely used horn is the one which is flared in both directions. It is widely referred to as a *Pyramidal Horn*, and its radiation characteristics are essentially a combination of the E- and H-plane sectoral horns. [1] And this is exactly why it is irrelevant that we don't have two other shafts' openings on the other sides of that horizontal cross section; whether the GPG receive pulses from an electric field or a magnetic one, the proportions are fixed and what matters is the wave length itself. It is important though not to confuse the 33π distance as the length of the Antenna, but rather as a significant constant which delivers the needed Image Resolution Factor.

What is astounding here is that if we to examine the directivity (aka Gain, for maximum power efficiency) of the Pyramidal Horn, we find that

$$\frac{4\pi A_e}{\lambda^2} = \frac{b_1 a_1}{\lambda^2}\frac{32}{\pi}$$

$$\frac{b_1 a_1}{A_e} = \frac{physical\ aperture}{effective\ aperture}$$

$$= 1.2337 = \sqrt[k]{68 \times 10{,}000}$$

Where

$$k = 64, 2^6, 8^2, (\frac{32.5}{32} - 1)^{-1}, 1.68^8$$

The last two numbers above are surprisingly significant and tell much about the relation between the GPG, Pyramidal Horn Antenna model and some celestial reference which will be presented shortly.

An *isotropic Antenna* is an ideal antenna that radiates its power uniformly in all directions. There is however no actual physical isotropic antenna. It is often used as a reference antenna for calculating the gain. [2] And by examining the dimensions of an isotropic Antenna, we find from reference [3] that

$$A_e = \frac{\lambda^2}{4\pi} = 0.0796 \times \lambda^2$$

And miraculously, when substituting $A_e = 33\pi/1.2337$, we get

$$\lambda = 32.5$$

Which is "my" number manifesting itself as a wavelength here!

Not just any length, but the most important length in the Antenna model that I am able to link to GPG numerically (*More information will be presented in the next Chapter*). Notice also that we have a one unit area of

$$\frac{1}{0.0796^2} = 1.57 \times 100 = \frac{\pi}{2} \times 100$$

The culmination of dimensions' disclosure here is in this specific equation

$$1\ steradian = \left(\left(\frac{\pi}{2} - 1\right) \times 100\right)^2$$

$$= 3282.80635\ square\ degrees$$

$$= (0.572 \times 100)^2$$

Where 1 radian equals to

$$\frac{180°}{\pi} = 57.2°$$

And 0.572 radians equal to

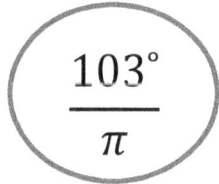

$$\frac{103°}{\pi}$$

The link to *32.5 System* starts with

$$\frac{325}{\pi} = 103$$

But more on that in the next chapter. Meanwhile, we also notice that the *angular size* of a celestial object viewed from Earth is

$$57 \times \frac{actual\ size}{distance}$$

$$\approx \left(\frac{\pi}{2} - 1\right) \times 100 \times \frac{actual\ size}{distance}$$

$$\approx \left(\frac{\pi}{2} - 1\right) \times 100 \times \frac{1{,}392}{149{,}000}$$

$$\approx 0.53325° = 32\ arcminutes$$

This is the angular size of the Sun; which also equals to

$$\approx \frac{1.68}{\pi}$$

This is incredible indeed, but that's not the whole story yet! The Sun as you can see is important in setting the right proportions, while the Moon is [T]he object that plays the major role. To get a true representation of the sizes, view the image below at a distance of 103 times the width of the "Moon: max." circle. For example, if this circle is 10 cm wide on your monitor, view it from 10.3 meters away.

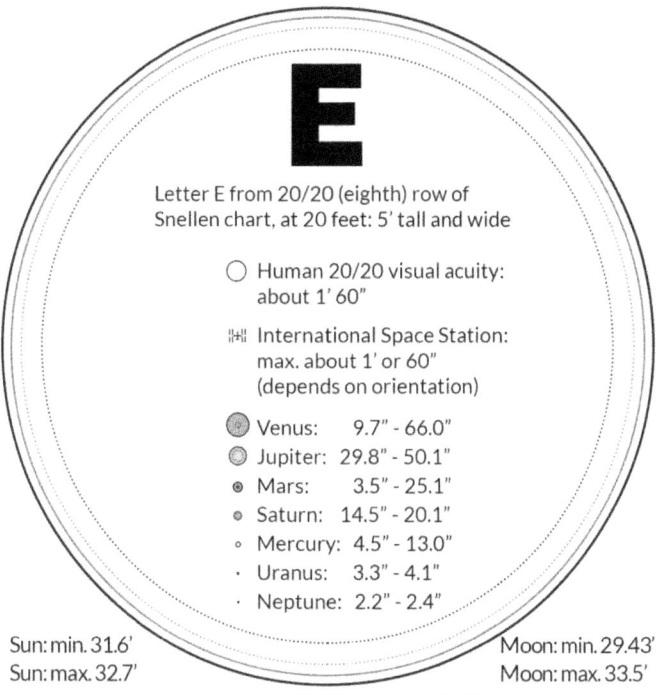

E

Letter E from 20/20 (eighth) row of
Snellen chart, at 20 feet: 5' tall and wide

○ Human 20/20 visual acuity:
about 1' 60"

▥ International Space Station:
max. about 1' or 60"
(depends on orientation)

◉ Venus: 9.7" - 66.0"
◎ Jupiter: 29.8" - 50.1"
◉ Mars: 3.5" - 25.1"
◦ Saturn: 14.5" - 20.1"
◦ Mercury: 4.5" - 13.0"
· Uranus: 3.3" - 4.1"
· Neptune: 2.2" - 2.4"

Sun: min. 31.6'
Sun: max. 32.7'

Moon: min. 29.43'
Moon: max. 33.5'

Figure 10: Reference [4][5]

THE RELEVANCE OF 103 WITH THE NUMBERS FROM THE BOOK '32.5 SYSTEM'

$$\frac{33.5}{32.5} = \frac{103}{100}$$

BESSEL FUNCTIONS

Bessel's equation arises when finding separable solutions to
Laplace's equation and the Helmholtz equation in cylindrical or
spherical coordinates. Bessel functions are therefore especially
important for many problems of wave propagation and static
potentials. In solving problems in cylindrical coordinate systems,
one obtains Bessel functions of integer order (α = n); in
spherical problems, one obtains half-integer orders (α = n+1/2).
[6]

Type	First kind	Second kind
Bessel functions	J_α	Y_α
Modified Bessel functions	I_α	K_α
Hankel functions	$H_\alpha^{(1)} = J_\alpha + iY_\alpha$	$H_\alpha^{(2)} = J_\alpha - iY_\alpha$
Spherical Bessel functions	j_n	y_n
Spherical Hankel functions	$h_n^{(1)} = j_n + iy_n$	$h_n^{(2)} = j_n - iy_n$

The interest in the table above lies in the basic Bessel functions
of J_α and Y_α since they represent the basic foundation for the
others as you can see. As mentioned earlier, the large circular
loop antenna modeling ($a \geq \lambda/2$) is tweaked using a very
relevant and interesting number to us. And to understand how
this number came about, we need to discern the equations that
delivered it to us, but since this book is not concerned with the
mathematical proof thereof, I will just present the relevant
equations to demonstrate the application of that number.

$$P_{\text{rad}} \simeq \frac{\pi(a\omega\mu)^2|I_0|^2}{4\eta(ka)}$$

The maximum radiation intensity occurs when $ka\sin\theta = 1.84$ so that

$$U|_{\text{max}} = \frac{(a\omega\mu)^2|I_0|^2}{8\eta}J_1^2(ka\sin\theta)|_{ka\sin\theta=1.84} = \frac{(a\omega\mu)^2|I_0|^2}{8\eta}(0.582)^2$$

Thus

$$R_r = \frac{2P_{\text{rad}}}{|I_0|^2} = \frac{2\pi(a\omega\mu)^2}{4\eta(ka)} = \eta\left(\frac{\pi}{2}\right)ka = 60\pi^2(ka) = 60\pi^2\left(\frac{C}{\lambda}\right)$$

$$D_0 = 4\pi\frac{U_{\text{max}}}{P_{\text{rad}}} = 4\pi\frac{ka(0.582)^2}{2\pi} = 2ka(0.582)^2 = 0.677\left(\frac{C}{\lambda}\right)$$

$$A_{em} = \frac{\lambda^2}{4\pi}D_0 = \frac{\lambda^2}{4\pi}\left[0.677\left(\frac{C}{\lambda}\right)\right] = 5.39 \times 10^{-2}\lambda C$$

where $C(\text{circumference}) = 2\pi a$ and $\eta \simeq 120\pi$.

The equations above were taken from reference [7]. We notice therefrom that the significant number of 0.677 is linked with the other number of 0.582. We are not able to directly identify the meaning of this second number showing up here; but with a slight modification to tune 0.677 into its rounded-off value of 0.68 (*which is presented by the other reference I mentioned earlier*), we find that

$$\frac{0.582^2}{0.677} = 0.5$$

$$\approx \frac{0.585^2}{0.68}$$

(*with* 0.5% *error only*)

This leaves the ratio equals to the same constant and remain in the proper range of error while linking both numbers to one another; the significance of the second number is mentioned in

32.5 System and it is also used in the next chapter. Also note from the table in *Appendix II* that the range (0.582~0.585) is to be located exactly in the same location, i.e. row, and nowhere else.

The real mystery which I am about to disclose now is yet more astounding still. For that from the Speed of Light value as we have seen it linked with the Royal Cubits up to a 10^{-8} precision before, we also have

$$\sqrt{2.99792458}^{\ -1} = 0.577$$

The value of 0.577 (*See Appendix IV*) shows up in Bessel functions and is called the Euler or Mascheroni constant. Although it shows up in the equation for the Bessel function of the second kind, but it accompanies the first kind term therein. [8] Which means that it is contributing to its magnification.

$$Y_n(x) = \frac{2}{\pi} J_n(x)\left(\ln\frac{x}{2} + \gamma\right) - \frac{1}{\pi}\sum_{k=0}^{n-1}\frac{(n-k-1)!}{k!}\left(\frac{2}{x}\right)^{n-2k}$$
$$- \frac{1}{\pi}\sum_{k=0}^{\infty}\frac{(-1)^k}{k!(n+k)!}[h_k + h_{k+n}]\left(\frac{x}{2}\right)^{n+2k}$$

where

$$h_k = 1 + \frac{1}{2} + \cdots + \frac{1}{k}$$

and

$$\gamma = \lim_{k\to\infty}\left(1 + \frac{1}{2} + \cdots + \frac{1}{k} - \ln k\right) = 0.577215\ldots$$

This works just fine down to a certain number of terms in *gamma*, but it serves well our application and there is no need for any further precision beyond the first three fractions of the decimal digits. This constant is basic in antenna design and also shows up for finite length dipole and microstrip antennas.

CHAPTER REFERENCES

[1] Antenna Theory: Analysis and Design, 3 rd Edition by Constantine A. Balanis.

[2] http://www.telecomabc.com/i/isotropic.html

[3] p. 29, Antennas: For All Applications, J. D. Kraus, R. J. Marhefka.

[4] https://en.wikipedia.org/wiki/Angular_diameter

[5] https://en.wikipedia.org/wiki/Minute_and_second_of_arc

[6] https://en.wikipedia.org/wiki/Bessel_function

[7] See reference [1].

[8] p. 121, Fourier Series, Transforms, and Boundary Value Problems, J. R. Hanna, J. H. Rowland.

5 THE LINK TO '32.5 SYSTEM'

From the article, we have seen the significance of 100π. And if we tune the equation in proper manner, we get

$$100\pi = \frac{44^2 \times 100}{2 \times 280 \times 1.1} = \frac{1.57}{2} \times \frac{440}{1.1}$$

$$= \frac{\pi}{2} \underbrace{(\sqrt{2} \times 10)^2}_{relevant}$$

This corresponds to the relation that we have in *32.5 System*:

$$celestial\ relation\ A = \frac{1}{27.32} + \frac{1}{29.25}$$

$$= \frac{1}{10 \times \sqrt{2}}$$

The significance of $\sqrt{2}$ lies in

$$\frac{440}{\sqrt{2 \times 440^2}} \times \pi = \sin \emptyset \times \pi$$

$$= \cos \emptyset \times \pi = \frac{\pi}{\sqrt{2}}$$

This equals to

$$\underbrace{\sqrt{2 \times 440^2}}_{diagonal\ unit\ length}^{-1} \times \underbrace{2 \times 220 \times \pi}_{circumference}$$

$$= \frac{\pi}{\sqrt{2}}$$

And hence, it is directly linked to the circumference of the inner circle (see the article) of the pyramid's base. It is also demonstrated in

$$\frac{outer\ circle\ radius}{inner\ circle\ radius} = \sqrt{2}$$

To complement the story here, we notice that

$$\frac{celestial\ relation\ A \times 10}{27.32}$$

$$= \frac{DUL \times CIRCUM}{\pi}$$

$$\rightarrow \frac{celestial\ relation\ A \times 366\pi}{1,000}$$

$$= DUL \times CIRCUM$$

But most importantly

$$\Delta = r_2 - r_1 = (\sqrt{2} - 1)\ r_1$$

Where $(\sqrt{2} - 1)$ is a factor that plays a major role celestially as well as geometrically in the GPG; celestially, it gives us:

$$celestial\ relation\ B = \frac{1}{27.32} - \frac{1}{29.25}$$

$$= \frac{1}{\sqrt{2} - 1} \times \frac{1}{1,000}$$

And geometrically (with only 0.4% error):

$$\frac{(perimeter = 440 \times 4)^2}{1,000,000} \times 33\ \ + 1$$

$$= 103\ RC$$

This means that for a radius of 1,000 we get a $\sqrt{circle\ area}$ of

$$1,000\sqrt{\pi}$$

And a total area of $(1,000,000\ \pi)$; turning the above equation into

$$\frac{(perimeter = 440 \times 4)^2}{\dfrac{circle\ area}{\pi}} \times 33\ + 1$$

$$= 103\ RC$$

$$\rightarrow \frac{perimeter^2}{circle\ area} \times 33\pi\ + 1 = 103\ RC$$

This is an incredible relation that did not come about by coincidence! We have seen so far (in the previous chapter) how

$$\frac{325}{\pi} = 103$$

But it is clear that we also have the following relation

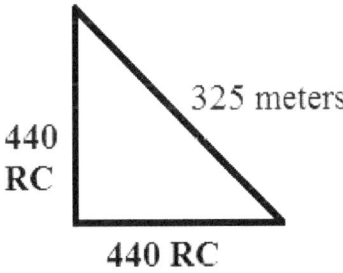

Going back to the circle's area; if we to extend these dimensions onto a square as in *32.5 System* (p. 225-230), we find that

$$5.85 = (celestial\ relation\ A \\ \times celestial\ relation\ B \\ \times 10)^{-1}$$

$$width \times length \times 10^{-6} \\ = square\ area \times 10^{-6} \\ = 5.85$$

THE LINK WITH THE SOLAR CALENDAR DRIFT

The significance of the number 10.88 was pointed out in *32.5 System* and now we can observe the relation of this number with the GPG's dimensions in the following

$$\frac{base\ length}{height} = \frac{440}{280} = 1.57$$

And

$$1.57 \cong \left(\frac{10.88}{10}\right)^{\frac{16}{3}}$$

The interesting thing about this relation is that if we keep the GPG's proportions fixed unto the 100π constant (as explained in the article) while magnifying the vertical cross-sectional area ($0.5 \times height \times width$) by a 1,000 factor, we'll have this ratio between the height and width

$$\frac{x^{\frac{2}{3}}}{x^{\frac{1}{3}}} = x^{\frac{1}{3}}$$

This means:

1. If the area is magnified by a 1,000 factor, the height becomes larger than the width by a factor of 10.
2. If the area is magnified by a 1,000,000 factor, the height becomes larger than the width by a factor of 100.

But miraculously, we have this:

$$\frac{100}{68.3} = \pi^{\frac{1}{3}}$$

And this:

$$x^{\frac{2}{3}} = 1.57 \rightarrow x = \left(\frac{10.88}{10}\right)^{8}$$

The latter delivers a divisor of 100,000,000 which is exactly relevant to the discovery I made on 27 April 2016.

6 NUMBER THEORY ANCHOR

T he whole discipline of numbers acts as an anchor serving as a central cohesive source of support and stability to the entire application that uses it. And this is what number theory does; it is the branch of mathematics that deals with the properties and relationships of numbers. The next phase of this book is concerned with inspecting any relations therein that could help better understand the structure of the GPG.

An interesting application in mathematics is the Basel problem. It is a problem in mathematical analysis with vivid relevance to number theory, first posed by Pietro Mengoli in 1644 and solved by Leonhard Euler in 1734. The Basel problem asks for the precise summation of the reciprocals of the squares of the natural numbers, i.e. the precise sum of the infinite series: [1]

$$\sum_{n=1}^{\infty} \frac{1}{n^2} = \lim_{n \to \infty} \left(\frac{1}{1^2} + \frac{1}{2^2} + \cdots + \frac{1}{n^2} \right)$$

The sum of the series is approximately equal to 1.644934; Euler found the exact sum to be [2]

$$\frac{\pi^2}{6}$$

and announced this discovery in 1735. [3] The infinite series whose terms are the natural numbers $1 + 2 + 3 + 4 + \cdots$ is a divergent series. The nth partial sum of the series is the triangular number [4]

$$\sum_{k=1}^{n} k = \frac{n(n+1)}{2}$$

To understand better the sum of such series, we look at the following surprisingly cross-sectional Pyramids:

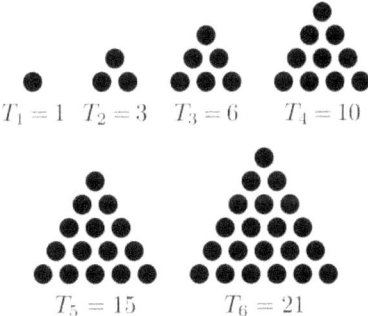

The first six triangular numbers [5]

Which increases without bound as n goes to infinity. Because the sequence of partial sums fails to converge to a finite limit, the series does not have a sum. Although the series seems at first sight not to have any meaningful value at all, it can be manipulated to yield a number of mathematically interesting results, some of which have applications in other fields such as complex analysis, quantum field theory, and string theory. Many summation methods are used in mathematics to assign numerical values even to a divergent series. In particular, the methods of zeta function regularization and Ramanujan summation assign the series a value of $-1/12$, which is expressed by a famous formula [6]

$$1 + 2 + 3 + 4 + \cdots + \infty = -\frac{1}{12}$$

Note that this equation is almost exclusively used in String Theory among all other scientific fields; one barely finds any other application for it! The Riemann zeta function or Euler–Riemann zeta function, $\zeta(s)$, is a function of a complex variable, s, that analytically continues the sum of the infinite series [7]

$$\zeta(s) = \sum_{n=1}^{\infty} \frac{1}{n^s}$$

which converges when the real part of s is greater than 1. The Riemann zeta function plays a pivotal role in analytic number theory and has applications in physics, probability theory, and applied statistics.

As a function of a real variable, Leonhard Euler first introduced and studied it in the first half of the eighteenth century without using complex analysis, which was not available at the time. Bernhard Riemann's 1859 article "On the Number of Primes Less Than a Given Magnitude" extended the Euler definition to a complex variable, proved its meromorphic continuation and functional equation, and established a relation between its zeros and the distribution of prime numbers.

The values of the Riemann zeta function at even positive integers were computed by Euler. The first of them, $\zeta(2)$, provides a solution to the Basel problem (*referred to in the beginning of this chapter*). In 1979 Apéry proved the irrationality of ζ(3). The values at negative integer points, also found by Euler, are rational numbers and play an important role in the theory of modular forms. Many generalizations of the Riemann zeta function, such as Dirichlet series, Dirichlet L-functions and L-functions, are known. [8]

$$\zeta(1) = 1 + \frac{1}{2} + \frac{1}{3} + \cdots = \infty$$

ENCODED NUMBER IN NUMBER THEORY

If we approach from numbers larger than 1, then this is the harmonic series. But its Cauchy principal value [9]

$$\lim_{\varepsilon \to 0} \frac{\zeta(1 + \varepsilon) + \zeta(1 - \varepsilon)}{2}$$

exists which is the (earlier mentioned) Euler–Mascheroni constant [10]

$$\gamma = 0.577$$

Three fractional decimal digits are enough for our application/framework. The reader should not view this as *cherry picking* since mathematicians and scientists deal with precisions in the same way. What matters is the usage which the number serves and not the precision with which the number was originally constructed. For example, from [11] one reads:

> *"so the answer is correct to three decimal places at least ...*
> *Therefore the calculations will be carried out to five places with*
> *the intention of retaining three places in the final answer."*

ENCODED MESSAGE IN NUMBER THEORY

An important relation starts to twinkle from the equation

$$1 + 2 + 3 + 4 + \cdots + \infty = -\frac{1}{12}$$

and the equation

$$\sum_{k=1}^{n} k = \frac{n\,(n+1)}{2}$$

as soon as we make them equal to one another; $n = \infty$:

$$1 + 2 + 3 + \cdots + n = \frac{n\,(n+1)}{2}$$

$$\rightarrow \quad -\frac{1}{12} = \frac{n\,(n+1)}{2}$$

$$\rightarrow \quad n^2 + n + \frac{1}{6} = 0$$

This delivers

$$n_1 = \frac{1}{2\sqrt{3}} - \frac{1}{2} = -0.211324$$

$$= -\frac{1}{(501.423)^{0.25}}$$

$$= -\frac{1}{\left(354.37\sqrt{2}\right)^{0.25}}$$

$$= -(0.678)^4$$

And

$$n_2 = -\frac{1}{2\sqrt{3}} - \frac{1}{2} = -\frac{1}{2}(1 + 0.577)$$

$$= -\frac{1}{2}(0.618 \times 10)^{0.25}$$

$$\approx -\frac{1}{2}\left(\frac{1}{17} + 1\right)^8$$

$$= -(1.2 + \frac{0.68}{10})^{-1}$$

(see Appendix IV for the relevance of the number 309 from the book of 32.5 System)

Now, and after I have calculated the values for both of n_1 and n_2, we observe the following

$$\sum_{k=1}^{n} k^2 = \frac{n(n+1)(2n+1)}{6}$$

and solving for, n, in

$$-\frac{1}{2} = n(n+1)(2n+1)$$

delivers three solution; two complex values (which we can ignore) and one single real number which equals to

$$n = -1.26 \approx 1/n_2$$

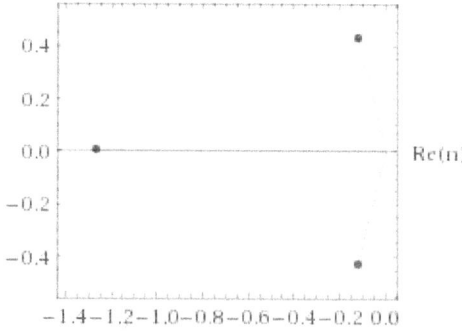

[Graph was generated on wolframalpha.com]

And yet for using

$$\sum_{k=1}^{n} k^3 = \frac{n^2 (n + 1)^2}{4}$$

there exist only complex solutions and no real ones as shown in the graph below.

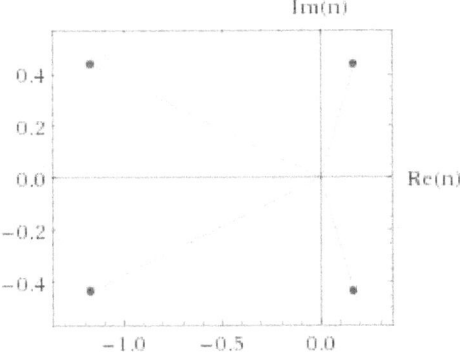

[Graph was generated on wolframalpha.com]

It is however known that

$$\sum_{k=1}^{n} k^3 = (\sum_{k=1}^{n} k)^2$$

Looking back at the triangular number ($T_3 = 6$) which were mentioned earlier, we find that

$$(1 + 2 + 3)^2 = 6^2 = 36$$

This equals to the wavelength which can be retrieved from the equation of

$$A_e = \frac{\lambda^2}{4\pi} = 0.0796 \times \lambda^2$$

if the effective aperture equals to 33π. In other words, we can model the celestial motions using these three numbers only to produce the following

$$1 + \frac{1}{2} + \frac{1}{3} = 1 + 0.8333$$

$$1 - \frac{1}{2} - \frac{1}{3} = 1 - 0.8333 = \frac{1}{6}$$

The significance of 0.8333 is that 10 months are 0.8333 of a year:

$$0.8333 = \frac{10}{12} = \frac{10,000}{12,000}$$

$$= \frac{1}{1.2}$$

Please refer to 32.5 System (p. 212) for further insights
(See Appendix IV)

THE SOLAR CALENDAR DRIFT AND THE PIVOTAL ROLE OF 32.5

looking at the following equation

$$1 + 2 + 3 + \cdots + n = \frac{1088.75}{2}$$
$$= \frac{33.5 \times 32.5}{2}$$

and at

$$\frac{1088.75}{2} - 32.5 + 31.5 + \cdots + 18.5$$
$$= 1.61875 \times 100$$
$$= \underbrace{17.5}_{corner} + 16.5 + \cdots + 1.5$$
$$+ 0.375$$

$$= \frac{325 - 1.25}{2} = 1.625 \times 100 - 0.625$$
$$= 1.61875 \times 100$$

one notices that:

$$17.5 = 0.057^{-1} = \frac{10}{0.57}$$
$$= 10 \times (\frac{\pi}{2} - 1)^{-1}$$

Which points to the role that 0.57 serves here as a corner for setting the length on the right proportion, but which proportion am I referring to exactly? If we to further inspect the above relation, we find that:

$$n_1 = \frac{\pi}{10.88} - \frac{1}{2} \qquad n_2 = -\frac{\pi}{10.88} - \frac{1}{2}$$

(See Appendix IV)

Impressive acrobatic is this 10.88 doing before our eyes! But what is quite astounding as well is when investigating the quantity of numbers, we find that

$$celestial\ reference$$
$$\underbrace{-\ 32.5 + \cdots + 18.5}_{15\ numbers}$$
$$= \underbrace{17.5 + \cdots + 1.5 + 0.375}_{18\ numbers}$$

$$\frac{15}{18} = 0.8333$$

To understand the significance of the celestial reference refer to
32.5 System
(See also Appendix V)

Eight minutes is the time it takes light to reach Earth from the Sun, and 0.8333 resembles the time needed to cross one tenth of that distance.

CHAPTER REFERENCES

[1] https://en.wikipedia.org/wiki/Basel_problem
[2] Ibid.
[3] Ibid.
[4]
https://en.wikipedia.org/wiki/1_%2B_2_%2B_3_%2B_4_%2B_
%E2%8B%AF
[5] Ibid.
[6] Ibid.
[7] https://en.wikipedia.org/wiki/Riemann_zeta_function
[8] Ibid.
[9] Ibid.
[10] Ibid.
[11] Riemann's Zeta Function, by Harold M. Edwards.

7 THE ANCHORAGE OF THE VERNAL EQUINOX

(This chapter is still in draft phase)

A stonishingly enough, the height of the Great Pyramid of Giza is so relevant in calculating the date on which the Vernal Equinox takes place; the following equation demonstrates this relation:

$$Distance\ from\ the\ Sun$$

$$= \frac{100}{\sqrt{Khufu's\ Height}} \times c_1 \times 60$$

where Khufu's height is in units of meters; and c_1 (m/sec) is the speed of light value. The distance obtained lies between the maximum distance from the Sun: 1.017 AU=1.521×10^8 km, and the minimum distance from the Sun: 0.983 AU=1.471×10^8 km. This coincides with the time period of March Equinox which is also known as the "Spring (aka, Vernal) Equinox" in the Northern Hemisphere and as the "Autumnal (aka, Fall) Equinox" in the Southern Hemisphere. The event was and still is significant for many civilizations and cultures worldwide and throughout history inasmuch as that one of the most famous ancient Spring Equinox celebrations was the Mayan sacrificial ritual by the main pyramid at Chichen Itza, Mexico. To understand the above equation better, one can inspect its separate elements first; for example, the speed of light value can be used either in the units of (meters per second) as it was substituted in the equation above, or in the units of (meters per minute or even per hour).

$$c_1 = c_2 \times 60 = c_3 \times 3600$$

With the units

$$c_1 \left(\frac{meter}{second}\right); \; c_2 \left(\frac{meter}{minute}\right); \; c_3 \left(\frac{meter}{hour}\right)$$

Integrating that back into the equation after squaring both of its sides, delivers the following:

$$(Distance \; from \; the \; Sun)^2$$

$$= \frac{10,000}{Khufu's \; Height} \times (c_1 \times 60)^2$$

The speed of light squared value in the equation above has the units of

$$c_1{}^2 \left(\frac{m^2}{sec^2}\right) = c_2{}^2 \left(\frac{m^2}{min^2}\right) \times 60^2$$

$$= c_3{}^2 \left(\frac{m^2}{hr^2}\right) \times 3600^2$$

And substituting with the desired units, we get:

$$(Distance\ from\ the\ Sun)^2$$

$$= \frac{10,000}{Khufu's\ Height} \times c_1{}^2 \times 3600$$

$$= \frac{10,000}{Khufu's\ Height} \times (c_2 \times 3600)^2$$

$$= \frac{10,000}{Khufu's\ Height} \times c_2{}^2 \times 3600^2$$

$$\rightarrow \sqrt{Distance\ from\ the\ Sun}$$

$$= \frac{10}{\sqrt[4]{Khufu's\ Height}} \times \sqrt{c_2} \times 60$$

$$\rightarrow \sqrt{\dfrac{Distance\ from\ the\ Sun}{c_2}}$$

$$= \dfrac{600}{\sqrt[4]{Khufu's\ Height}}$$

$$= \dfrac{65 \times \dfrac{300}{32.5}}{\sqrt[4]{Khufu's\ Height}}$$

$$= \dfrac{60 \times \dfrac{325}{32.5}}{\sqrt[4]{Khufu's\ Height}}$$

ANCHORING THE EXACT DATE OF THE CALENDAR ONTO WHICH THE VERNAL EQUINOX FALLS

In the provided tables from NASA *(See Appendix VII)*, we can exactly determine on which day the Vernal Equinox takes place.

it falls between DOY 69-72 if we consider the GPG height to equal 146.6, because 1.4856/1.496=0.993
even if you don't round it off, the drift is only in one single day DOY 68 which is still in the range and in March.
It can be calculated though to account for the leap year's extra/less day. It provides us with that specific precision. Which is awesome!

85

If we want the equinox to fall on 20th/21st of March, the height of the pyramid will be 145.6 meters. One meter less! (Taking into consideration that the table provided by NASA is valid always!) ..

this could also explain the base under the pyramid, to account for the one missing meter to push forth the day of celebration about 10 days! A correction measure as in the Roman calendar guys ..

Christopher Knight & Alan Butler speak of 279 Cubits height in their book "Civilization One".

The DOY works hand in hand with the measure of 10,000 in that 1/146.6 of the GPG's length signifies the equinox start, i.e., 68.

$$146.6 = \frac{10,000}{68.2}$$

The ancient Egyptians encountered the same miscalculation in the Great Pyramid as that which the Romans faced in their Julian Calendar; 10 days drift away from the 20th/21st of March (i.e., Equinox day). The Romans amended the calendar into the Gregorian system, and the ancient Egyptians tried to correct the GPG's height using the capstone and the base underneath the Great Pyramid.

The Great Pyramid of Giza were meant to express the Vernal Equinox through its dimensions, not the whole calendar.

Khafre's height difference from that of Khufu's represents a distance of ten multiples (closer to)/(away from) the Sun in reference to Spring Equinox Day on Earth!!

Notice that this value equals to 1,000/(GPG height)

$$\frac{F(146.608 - 143.5)}{1.49597} = \frac{68.2}{10} \; Astronomical \; Units$$

$$F = \frac{100}{\sqrt{height}} \times speed \; of \; light \times 60$$

(http://landsathandbook.gsfc.nasa.gov/excel_docs/d.xls) from day 57 to day 93 we have 0.99. This is where we have the right precision to state that: 6.82/0.99=6.8. Whereas, 6.82/0.98=6.9. In other words, the whole month of March is included therein to guarantee that the whole window of opportunity is therein for March, but no other months.

And Menkaure's pyramid equation goes like this:
F(143.5-65=68.5+10)=146(200)/(146^2+200)

Don't be confused by the units of "square arcminutes". These refer to surface area of a sphere with radius of 1 meter.

GPG's encoded distance from the Sun in reference to the

Vernal Equinox

$$= 1,000$$

× *the number of square arcminutes in a complete sphere*

p.s., the drift equals to (GPG height/10)^4. Which even has a meaning delivering a wonderful interpretation as well. 3*3.6=10.8= radius*pi/1000, where radius = 1 arcmin in radians), this means that 10.8/pi=10,800/(10*100pi)

$$(3.6 \times 3 \times 1,000)^2 = (360 \times 3 \times 100)^2$$

$$=$$

$$\frac{\pi}{4} \times number\ of\ arcminutes\ in\ a\ complete\ sphere$$

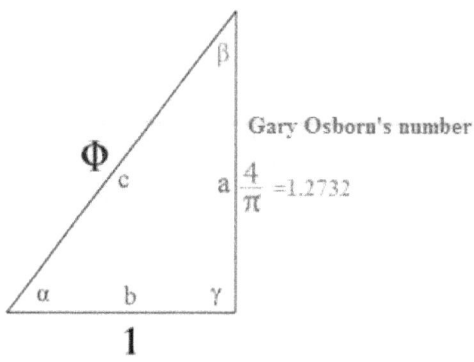

Gary Osborn's number

$$a\left|\frac{4}{\pi} = 1.2732\right.$$

NORTH SIDE BASE APOTHEM: 115.1456539393731 METRES,
DIVIDED BY THE HEIGHT OF 146.608 METRES = 0.7853981633974483 METRES ($\pi/4$).

a = | 1 METRE | 1

b = | 0.7853981633974483 METRES | $\frac{\pi}{4}$

c = | 1.2715542753135176 METRES | $\sqrt{\Phi}$

α = | 51.853974012777453° |

β = | 38.146025987222547° |

γ = | 90° |

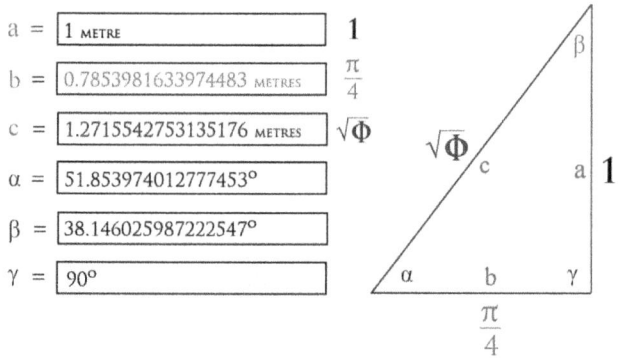

1) because, area=4*pi*r^2
2) and yet more precisely --> (33*pi/10*3+0.5236)

The significance of introducing a magnification factor of 1,000 to the square arcminutes sphere is to transfer the radius from being 1 meter to $1 \times \sqrt{1,000}$ meters.

$$\sqrt{1,000} \approx$$

$$\left(\underbrace{\frac{33 \times \pi}{10}}_{tenth\ of\ this\ significant\ length\ (meters)} \times 3 \right.$$

$$+ \underbrace{\frac{\sqrt{2} \times 440}{1,000}}_{thousandth\ of\ GPG's\ Base\ Diameter\ (meters,\ but\ project\ RC)} \left.\right)$$

1) Could it be that the Royal Cubits were devised for the purpose of complementing the square root of 1,000?
2) Could it be possible that the original idea was that a thousandth of the base diameter value should have served as a royal cubit? Because 523.6/sqrt(2)=440-69.758 and these 69.758 RCs=365.25/10 meters! (Which is one meter each 10 days on a tropical year calendar)

$$\frac{\sqrt{1,000}}{3} = \frac{33 \times \pi}{10} + \frac{1}{(\frac{\pi}{2} - 1) \times 10} = \frac{33 \times \pi}{10} + \frac{0.5236}{3}$$

$$\rightarrow \sqrt{1,000} = \frac{33 \times \pi}{10} \times 3 + 0.5236$$

Observe the FREQUENCY here. It is another indication of the GPG's design according to the solar calendar. In other words, this is an angular acceleration which were incorporated into the GPG's dimensions.

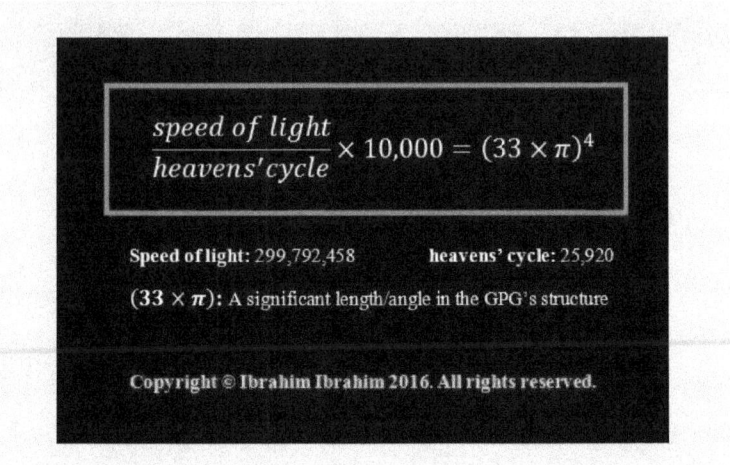

$$\frac{speed\ of\ light}{heavens'\ cycle} \times 10{,}000 = (33 \times \pi)^4$$

Speed of light: 299,792,458   heavens' cycle: 25,920

$(33 \times \pi)$: A significant length/angle in the GPG's structure

Note that this beautiful equation is telling us ALSO that dividing a length by 10 is crucial for some specific engineering application in the design of the GPG itself in relation with the heavens/light.

1/10th popped up in the measurements of the horizon from the GPG and the height of Khafre in reference to the Vernal Equinox and some other applications.

10,000/(heavens' cycle) refers to the age B.C. that Robert Bauval and Graham Hancock refer to!

And if we account for another 800 years (in case 10,800 is the officially established number of years), then the missing 800 contribute with this:

(32.5)*800/(heavens' cycle)=1, (it serves as a max margin value)

The one Royal Cubits addition on the length of the Pyramid's base side evokes the presence of such a relation:

$$\tan^{-1} \frac{1}{0.5236} \approx \frac{\sqrt{2} \times (440 + 1)}{10}$$

Celestial view according to terrestial angular acceleration were embedded in the structure of the Great Pyramid of Giza.

since tan 1/0.5236=1/30

Multiplying both sides by 100 reveals that both terms on the right side of the equation are in right proportion to one another. On one term the number '30' will show up and on the other, the number '10'; this proves that this equation serves the angles as much as the lengths!

$$\sqrt{1,000} \approx$$

$$\left(\underbrace{\frac{33 \times \pi}{10}}_{tenth\ of\ this\ significant\ length\ (meters)} \times 3 \right.$$

$$\left. + \underbrace{\frac{\sqrt{2} \times 440}{1,000}}_{thousandth\ of\ GPG's\ Base\ Diameter\ (meters,\ but\ project\ RC)} \right)$$

The multiplication with 100 means that:
100*sqrt(1,000)=pi*1,000

It seems highly probable to me that the unit of Royal Cubit -if it were defined based on its usage in the Pyramid's structure- were devised to account for one meter measure of length per ten days on a tropical year calendar.

This makes Menkaure refer to the speed of light in hours instead of seconds.

Moving from $\sqrt{1,000}$ to $100 \times \sqrt{1,000} = \pi \times 1,000$ delivers:

$$\frac{33 \times \pi}{10} \times 3 \times 100 = 33 \times \pi \times 30$$

So if we to multiply 30 with the following length, we get

$$\frac{100}{\sqrt{146.608}} \times 30 = 247.766°$$

$$= \frac{432}{100} \, rad = \frac{25,920}{60 \times 100} rad$$

Now we can write the equation in this form as well:

$$\frac{100}{\sqrt{146.608}} \times 30 \times 60 = \frac{25,920}{100} rad = \frac{100}{\sqrt{146.608}} \times \frac{speed\ of\ light}{\pi \times 1,000^{\,2}} \times 60$$

This resembles the earlier relation I devised in linking the GPG to the Vernal Equinox!

And for Menkaure:

$$\frac{100}{\sqrt{65}} \times 30 \approx 60 \times \frac{\sqrt{2} \times 440}{100} \ (degrees)$$

The Temporal Mark of GP is perpendicular: Vernal Equinox (celestial) - seconds.
The Spatial Mark of Khafre is parallel: Sphinx (looking East to sunrise) (celestial) - minutes.
The Angular Mark of Menkaure is rotational: The Offset (terrestrial) - hours.

For those civilizations which devised new ways to establish different bases of units of time (e.g., second, minute and hour), the aim was to control as much as possible their main festival date so that it falls on the exact Vernal Equinox day.

Correction: on the right hand side, solar year/(seconds per PY)

8 GENERAL RELATIVITY
(This chapter is still in draft phase)

It can be said that two objects in space orbiting each other in the absence of other forces are in free fall around each other. In general relativity, an object in free fall is subject to no force and is an inertial body moving along a geodesic. Far away from any sources of spacetime curvature, where spacetime is flat, the Newtonian theory of free fall agrees with general relativity but otherwise the two disagree. The experimental observation that all objects in free fall accelerate at the same rate, as noted by Galileo and then embodied in Newton's theory as the equality of gravitational and inertial masses, and later confirmed to high accuracy by modern forms of the Eötvös experiment, is the basis of the equivalence principle, from which basis Einstein's theory of general relativity initially took off. [1]

The existence of a mass will produce a curvature in space-time around it. Since light will follow the shortest path, or follows a geodesic of spacetime, then if the Earth curves the space around it then light passing the Earth will follow that curvature. [2] The distance of Earth from the Sun is about 108 times the diameter of the Sun (actually closer to 107.51, as per definition of the AU). Actual ratio varies between 105.7 (Perihelion) and 109.3 (Aphelion). [3]

$$d = \frac{1}{2}gt^2 \qquad\qquad t = \sqrt{\frac{2d}{g}}$$

d is the distance travelled by an object falling for a time t, and t is the time taken for an object to fall the distance d. In other words, we have the following proportional relation

$$\tau^2 \ \alpha \ distance$$

This corresponds to the Inverse-Square Law: [4]

$$distance^2 \ \alpha \ \frac{1}{acceleration}$$

The Arabian mythical horse can hence leap using the following parameters:

$$\rightarrow distance^2 = k \times \frac{time^2}{distance}$$

$$\frac{distance^2}{k}$$

$$= \left(\frac{43,200}{299792458^2} \right.$$

$$= \left. \frac{144}{299792458 \times 10^6} \right)$$

And since we have [5]

$$a = \frac{GM}{r^2} = \frac{k}{r^2}$$

Where

$$r = distance \ (m)$$

$$G = 6.672 \times 10^{-11} \ (m^3.kg^{-1}.s^{-2})$$

$$g = 9.8 \ (m.s^{-2})$$

$$M = 5.972 \times 10^{24} kg$$

After having substituted the acceleration value with the right parameters above, we now calculate the evaluated distance (r_e):

$$r_e = \sqrt{\frac{144}{299792458 \times 10^6 \times 5.972 \times 10^{24}} \times 9.8}$$

$$= 5302050.159 \ meters$$

Also note that

$$r_e = r \times \sqrt{\frac{g}{G}} = \sqrt{\frac{M \times g}{a}}$$

$$= r \times \sqrt{Great\ Pyramid's\ Height}$$
$$\times\ 31622.7766$$

$$\rightarrow r \times \sqrt{Great\ Pyramid's\ Height}$$
$$= \frac{r_e}{31622.7766}$$
$$= \frac{5302050.159}{31622.7766}$$
$$\approx 200 \times \theta\ (rad)$$

$$\rightarrow r \times \sqrt{\frac{Great\ Pyramid's\ Height}{40000}}$$

$$\approx r \times \sqrt{\frac{1}{\underbrace{27.32}_{days\ per\ sidereal\ month} \times 10}}$$

$$\approx \theta(rad)$$

$$r \approx \frac{\theta(rad)}{\sqrt{\frac{1}{\underbrace{27.32}_{days\ per\ sidereal\ month} \times 10}}}$$

$$\approx \frac{r_e \times \sqrt{273.2}}{L}$$

This demonstrates how the leap into the horizon (i.e., r) is proportional to Earth's mass and to the number of days per 10 sidereal months.

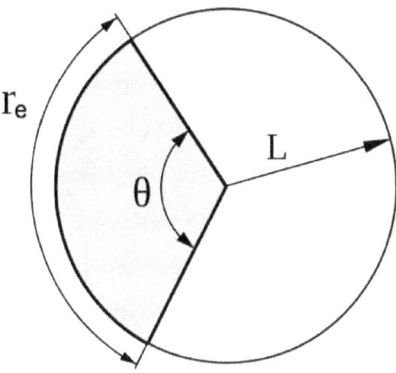

But we know the radius of the Earth (L) which equals to 6,371 km; hence, we can calculate the angle in the following:

$$\theta = \frac{r_e}{L} = \frac{5302050.159}{6371000}$$

$$= 0.8322 \left(= \frac{1}{1.2} \right) rad$$

$$= 47.68°$$

This is the evidence that seven and a half rotations around Earth in one second corresponds to the speed of light!

Conclusion 1: The Great Pyramid Of Giza Had Expressed The Horizon Through Its Dimensions In Correspondence With Planet Earth As A Whole.

CHAPTER REFERENCES

[1] https://en.wikipedia.org/wiki/Equations_for_a_falling_body
[2] Ibid.
[3] https://en.wikipedia.org/wiki/108_(number)
[4] https://math.uchicago.edu/~may/VIGRE/VIGRE2010/REUPaper s/Tolish.pdf
[5] Ibid.

9 THE GEOMETRY OF THE GIZA PLATEAU & THE SPEED OF LIGHT

(This chapter is still in draft phase)

$$\frac{Great\ Pyramid's\ Height}{40000}$$

$$\approx \frac{1}{27.32 \times 10}$$

$$\rightarrow \frac{Great\ Pyramid's\ Height}{200^2 = 10^8 \times (1 - \cos \alpha)^2}$$

$$\approx \frac{1}{27.32 \times 10}$$

$$where \quad \left(1 - \frac{2}{100}\right)$$

$$= \cos \alpha$$

$$= \cos \left(11.5° \times \frac{\pi}{180}\right)$$

$$and \quad \frac{11.5}{\pi} = \frac{366}{100} = \frac{100}{27.32}$$

$$and \quad \sqrt{speed\ of\ light} = 11.471^4$$

$$\approx 11.5^4$$

The speed of light value corresponds to a magnification power (of Menkaure's drift) to the factor of 8.

$$\rightarrow \frac{Great\ Pyramid's\ Height}{10^8 \times (1 - \cos \alpha)^2}$$

$$\approx \frac{1}{27.32 \times 10}$$

$$where \quad \frac{1}{\sqrt{\pi \times 2.99792458}} = \frac{325.8}{1,000}$$

$$\rightarrow \frac{Great\ Pyramid's\ Height}{10^8 \times (1 - \cos \alpha)^2 \times 2.99792458}$$

$$\approx \frac{\pi \times 325.8^2}{27.32 \times 10^7}$$

$$\rightarrow \frac{Great\ Pyramid's\ Height}{speed\ of\ light}$$

$$\approx \frac{\pi \times (1 - \cos \alpha)^2 \times 325.8^2}{27.32 \times 10^7}$$

$$\rightarrow \frac{GP\ Height}{speed\ of\ light \times GP\ Diagonal^2}$$

$$\approx \frac{\pi \times (1 - \cos \alpha)^2}{27.32 \times 10^7}$$

$$\rightarrow \frac{1}{speed\ of\ light \times 4\pi}$$

$$\approx \frac{181 \times (1 - \cos \alpha)^2}{27.32 \times 10^7}$$

$$\rightarrow \frac{1}{speed\ of\ light \times 4\pi}$$

$$\approx \frac{181 \times 4}{27.32 \times 10^{11}} \approx \frac{26.6}{10^{11}}$$

$$where \quad 6.672 \times 10^{-11}$$

$$\approx \frac{26.6}{10^{11} \times 3.98}$$

$$\rightarrow \frac{1}{speed\ of\ light \times 4\pi \times \underbrace{6.672 \times 10^{-11}}_{Gravitational\ Constant}}$$

$$\approx \frac{181(\approx 180) \times 4 \times 3.98}{27.32 \times 26.6} \approx 3.98 \approx 4$$

$$\therefore \quad \boldsymbol{4 \times speed\ of\ light \times 4\pi \times Gravit.\ Const. = 1}$$

Substituting it into:

$$a = \frac{GM}{r^2}$$

Gives

$$a = \frac{M(= 5.972 \times 10^{24})}{4c \times 4\pi \times r^2}$$

Substituting the radius of the Earth – which equals to 6,371 km– into r, we get:

$$a \approx 9.8 \ (m.\,s^{-2})$$

Notice also from the earlier equation

$$\rightarrow \frac{1}{speed\ of\ light \times 4\pi}$$
$$\approx \frac{57.29 \times \pi \times (1 - \cos \alpha)^2}{27.32 \times 10^7}$$

$$\rightarrow \frac{1}{speed\ of\ light \times 4\pi}$$
$$\approx \frac{2\pi \times (1 - \cos \alpha)^2}{10^7}$$

$$\rightarrow \frac{10^7}{2 \times speed\ of\ light} \approx [\ \underbrace{2\pi \times (1 - \cos \alpha)}_{\substack{solid\ angle\ of\ a\ spherical \\ cap\ of\ a\ unit\ sphere}}\ ]^2$$

$$\rightarrow \frac{10^7}{2} \approx [\underbrace{\frac{10^7}{\sqrt{\pi} \times 325.8} \times 2\pi}_{=10.88 \times 10,000} \times (1 - \cos \alpha)]^2$$

$$\rightarrow \frac{10^7}{2}$$

$$\approx [\ \underbrace{\frac{10^7}{\underbrace{\sqrt{\pi} \times 325.8}_{=r^2,\ \ r\ =360 \times 0.36524 \approx 360/2.732}} \times 2\pi \times (1 - \cos \alpha)}_{\substack{solid\ angle\ of\ a\ spherical \\ cap\ of\ a\ sphere\ with\ radius\ r}}\ ]^2$$

$$\rightarrow 1 \approx \underbrace{\sqrt{\frac{2 \times 10^7}{\pi}}}_{\substack{Giza\ Plateau \\ Normalization \\ Factor}} \times \underbrace{\frac{2\pi}{325.8}}_{\substack{Great\ Pyramid \\ Wavenumber}}$$

$$\times \underbrace{(1 - \cos\alpha)}_{\substack{The\ Giza\ Plateau \\ Angular\ Leakage}}$$

However, and as mentioned in Part I of this document, we know that seven and a half rotations around Earth in one second corresponds to the speed of light; hence:

$$\rightarrow 1\ m^2 = 1\ [Steradian.\, m^2] \approx$$

$$\underbrace{7.5}_{\substack{Giza\ Plateau \\ Reciprocal\ Wave\ Length \\ \sim 7.38904 = e^2 \\ \approx 7.3 = \sqrt[4]{\alpha} \times 4}} \times 2\pi \times \underbrace{(1 - \cos\alpha)}_{\substack{The\ Giza\ Plateau \\ Angular\ Leakage}}$$

Characteristic Equation of the Giza Plateau's Solid Angle

Half the value of the natural logarithm constant corresponds to an 8th root (of Menkaure's drift).

Great Pyramid Unit Diagonal: 1/325.8

Great Pyramid 'Wavenumber' or 'Phase Constant': 2π/325.8

Giza Plateau Normalization Factor: $\sqrt{(2 \times 10^7)/\pi}$

Giza Plateau Reciprocal Wave Length: $\sqrt{(2 \times 10^7)/\pi}$ */325.8 ≅ 7.5*

Giza Plateau 'Wavenumber' or 'Phase Constant': 7.5 × 2π

The Wavenumber is the number of radians per unit distance and the variation of α correspond to a precision only within 0.5° wherein the completion of seven and a half measure of light's rotations around Earth is attained.[†]

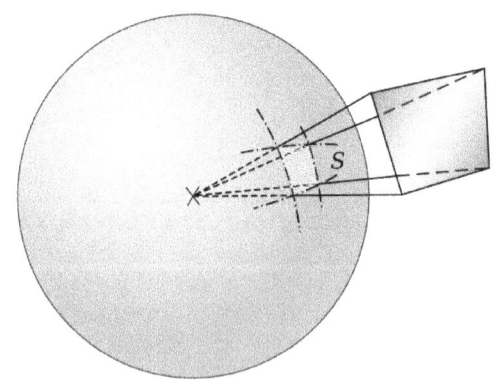

[†] Refer to Quote No. 72 in my book 'Quotable: My Worldview'.

From the figure above we see that the solid angle at the center of the sphere is subtended by the area surface S. And when this surface area equals to e^2 then the radius of that sphere also equals to the value of the *natural logarithm constant e* with a solid angle of $\Omega = 1$ and an angle of $\alpha = 32.77° = \frac{65.54°}{2} = \sqrt{10.88 \times 10 \times \pi^2}$. Notice however the following proportions in the figure below [1] [2] [3]

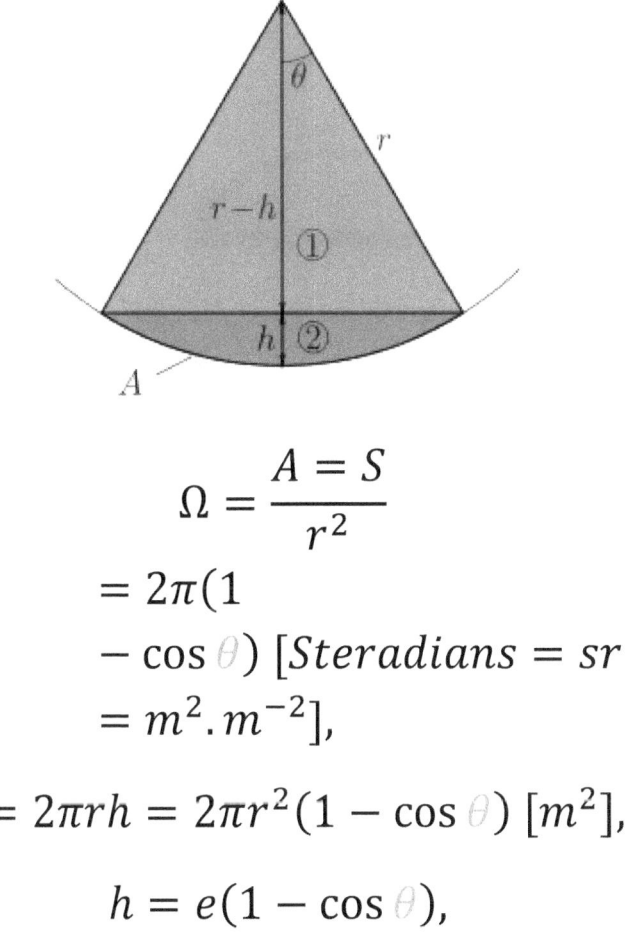

$$\Omega = \frac{A = S}{r^2}$$

$$= 2\pi(1 - \cos \theta) \; [Steradians = sr = m^2.m^{-2}],$$

$$A = 2\pi rh = 2\pi r^2 (1 - \cos \theta) \; [m^2],$$

$$h = e(1 - \cos \theta),$$

$$\frac{r - h}{r} = \cos \theta,$$

$$r = e = 2.71828 = \left(\frac{a^2 + h^2}{2h}\right),$$

a is the radius of the base of the cap

shorter

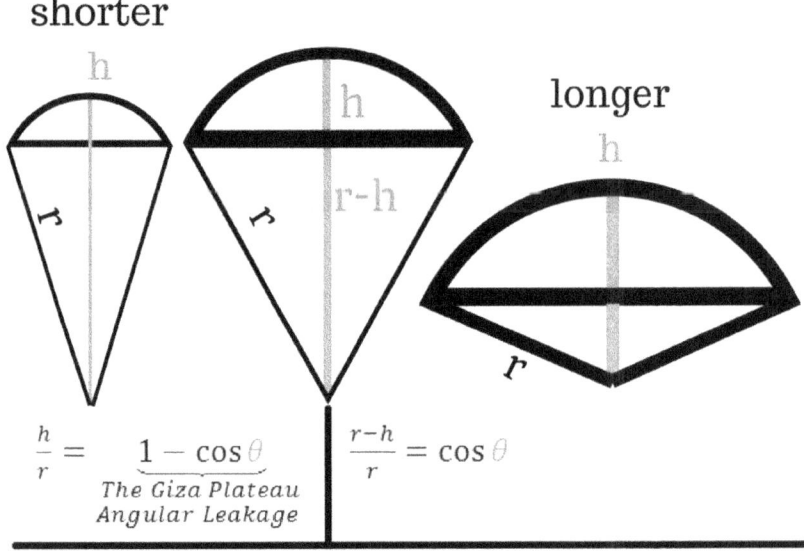

longer

$$\frac{h}{r} = \underbrace{\frac{1 - \cos \theta}{}}_{\text{The Giza Plateau}}$$
The Giza Plateau
Angular Leakage

$$\frac{r - h}{r} = \cos \theta$$

$$\rightarrow 1 \approx \frac{\text{Giza Plateau}}{\text{Normalization}} \times \frac{\text{Great Pyramid}}{\text{Wavenumber}} \times \frac{\text{The Giza Plateau}}{\text{Angular Leakage}}$$

- The *Characteristic Equation of the Giza Plateau's Solid Angle* reveals to us the Giza Plateau's Wavenumber where its Wave Length equals to the Great Pyramid's Normalized Base Diagonal and the Normalization Factor being used maps the Great Pyramid's Base Diagonal onto a circle's circumference with a radius which extends to an amount that equals to the GP's slope angle (**325. 8 = 2π × 51. 85**).

- It takes Earth 72 years to pass through 1° of the Zodiac, or 25,920 years to complete one full cycle; incredibly enough, this corresponds with the *Characteristic Equation Of The Giza Plateau's Solid Angle* as well.

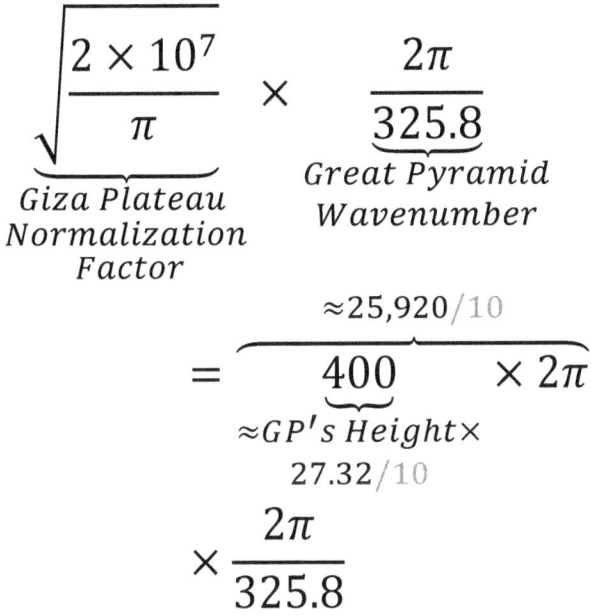

$$\underbrace{\sqrt{\frac{2 \times 10^7}{\pi}}}_{\substack{Giza\ Plateau \\ Normalization \\ Factor}} \times \underbrace{\frac{2\pi}{325.8}}_{\substack{Great\ Pyramid \\ Wavenumber}}$$

$$= \overbrace{\underbrace{400}_{\substack{\approx GP's\ Height \times \\ 27.32/10}}}^{\approx 25,920/10} \times 2\pi$$

$$\times \frac{2\pi}{325.8}$$

The Giza Plateau Normalization Factor equals to one tenth of the number of years needed for Earth to pass through 360 degrees of the zodiac and complete that one full cycle.

Inspecting the above equations further,

we find:

$$\theta = 11.5°, h = 17.6^{-1}, r = 2.81$$
$$\approx e \ (3\% \ error \ corresponding$$
$$with \ 0.5° \ variation)$$

For a complete sphere, we have:

$$\Omega = 4\pi = \frac{1}{r^2} \ Steradians$$

Making

$$r \approx \frac{100}{354.36}, \quad \theta = 180°$$

But in our case, we have

$$\Omega = \frac{1}{r^2} = \frac{1}{e^2} = \frac{1}{7.39} \ [sr]$$

And

$$A = \frac{r^2}{e^2} = 1$$

Which means that any portion of our sphere's surface (with area $A = 1$) subtends $\frac{1}{e^2}$ of a steradian; making thereby the amount of sections contained in a complete sphere equals to $\frac{4\pi}{e^2} = \frac{17}{10} = \frac{1}{10 \times h}$. Notice here that $\frac{17}{10} = (\frac{3.5436}{e})^2$, and if we to use the imprecise figure of 17.6 that we have seen earlier then the measure will change from 3.5436 into 3.60 marking thereby a shift from a lunar reference into a full-circle one!

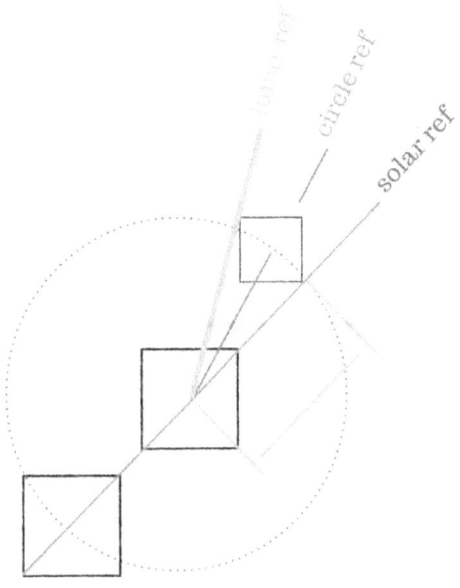

Such a full-circle reference can further be proven by simply invoking the lunar reference of 32.5 through the following relation

$$\frac{4 \times (3.25)}{e^2} = \frac{17.6}{10} = (\frac{3.6}{e})^2$$

$\rightarrow$ *leap reference marking a full circle*

In contrast to

$$\frac{4 \times (\pi)}{e^2} = \frac{17}{10} = (\frac{3.5436}{e})^2$$

$\rightarrow$ *leap circle marking the lunar year*

$$\frac{4 \times (3.35)}{e^2} = \frac{18}{10} = (\frac{3.66}{e})^2$$

$\rightarrow leap\ reference\ marking\ a\ 366\ circle$

This last equation is the result of obtaining

$\theta = 11.5°, h = 18^{-1}, r$

$= e\ (without\ any\ margin\ of\ error)$

Because for $A = 2\pi rh$, we have

$$A = 0.94 = 2(3.14) \times \frac{e}{18}$$

$$A = 0.98 = 2(3.25) \times \frac{e}{18}$$

$$A = 1 = 2(3.35) \times \frac{e}{18}$$

And

$$A = 0.97 = 2(3.14) \times \frac{e}{17.6}$$

$$A = 1 = 2(3.25) \times \frac{e}{17.6}$$

$$A = 1.03 = 2(3.35) \times \frac{e}{17.6}$$

And

$$A = 1 = 2(3.14) \times \frac{e}{17}$$

$$A = 1.03 = 2(3.25) \times \frac{e}{17}$$

$$A = 1.07 = 2(3.35) \times \frac{e}{17}$$

Conclusion 2: The Angular Drift Of The Pyramid Of Menkaure Marks The Whole Plateau Of Giza As Part Of One Single System.

CHAPTER REFERENCES

[1] https://en.wikipedia.org/wiki/Spherical_cap
[2] https://en.wikipedia.org/wiki/Solid_angle
[3] Atoms, Molecules and Optical Physics 1: Atoms and Spectroscopy, by Hertel, Ingolf Volker, Schulz, Claus-Peter, page 556.

10 THE MOON AND THE BASE OF THE GREAT PYRAMID OF GIZA
(This chapter is still in draft phase)

Unite Area = 16

GP Has Four Sides
With Four Parts
For Each Side

$$180 = 33\pi \times \underbrace{1.736}_{moon\ radius\ \times\ 10^{-6}} \times 16.5$$

$$= 2\pi \times 1.736 \times 16.5 = 2\pi \times 1.736 \times \frac{51.85°}{\pi}$$

$$\left(\sqrt{16.5} - \sqrt{16}\right) \times 440 = 1787.28 - 1760$$

$$= \underbrace{27.28}_{days\ in\ a\ sidereal\ month}$$

$$180 = 2\pi \times 1.736 \times \frac{51.85°}{\pi} = \underbrace{\frac{2 \times 1.736}{moon\ diameter \times 10^{-6}}} \times 51.85°$$

Multiplying the above equation with $\frac{\pi}{180}$ to convert the degrees into radians, we get

$$\pi = 2\pi \times 1.736 \times \frac{2 \times 12^2}{1,000}$$

$$\frac{1}{2} = Moon's\ Diameter \times \frac{12^2}{1,000} \times 10^{-6}$$

11 THE IMPEDANCE OF FREE SPACE & THE ANCIENT EGYPTIAN ROYAL CUBIT
(This chapter is still in draft phase)

The impedance of free space is a physical constant relating the magnitudes of the electric and magnetic fields of electromagnetic radiation travelling through free space. It also equals the product of the vacuum permeability and the speed of light in vacuum. [1] And now we notice that

$$\frac{360}{11.459} \times 12 = 377 \ (Impedance \ of \ Free \ Space)$$
$$\approx 366 + 10.88$$

$$\frac{360}{11.5} \times 12 = 375 \ (Impedance \ of \ Free \ Space)$$
$$\approx 365 + 10.88$$

And that[4]
$$11.459° = 0.2 \ rad$$

We also know that the angular velocity is the rate of change of angle with respect to time [2]

---

[4] To understand the significance of 11.459, refer to [3].

$$\omega = \frac{d(\varphi \, rad)}{dt} = \frac{d(l/r)}{dt} = \frac{c}{r} = \frac{299792458}{r}$$

$$r \times d(\varphi \, rad) = 299792458 \times dt$$

$$1 = \frac{299792458}{0.2 \times (Radius \; of \; Earth = 6,371,000)} \times T(sec)$$

Where

$$T^{-1} \approx \frac{440 \times 4}{7.5}$$

And

$$\frac{299792458}{7.5} \approx 6,371,000 \times 2\pi$$

Hence

$$\frac{T^{-1}}{7.5} \approx \frac{440 \times 4}{7.5^2} \approx \frac{377}{12}$$

$$T^{-1} = \mathbf{7.5} \left(sec^{-1}\right) \times \frac{377}{12} = \frac{\mathbf{299792458}}{\mathbf{6,371,000} \times 2\pi} \times \frac{377}{12}$$

$$= \frac{\mathbf{299792458}}{\mathbf{6,371,000}} \times \frac{60}{12} = \frac{\mathbf{299792458}}{\mathbf{6,371,000}} \times 5$$

But we have

$$T^{-1} + 47 = \frac{\mathbf{299792458}}{\mathbf{6,371,000}} \times 6 = 280 \; (Royal \; Cubits)$$

Delivering to us the following result

$$280 = \frac{440 \times 4}{7.5} + 47 \approx \frac{440 \times 4 + Days \; in \; a \; Lunar \; Year}{7.5}$$

(The error[5] is less than 1% and can further be tweaked)

What is more exciting though is the following

$$47\ RC = 23.5 \times 2 \approx 24meters \equiv 24\ hours$$

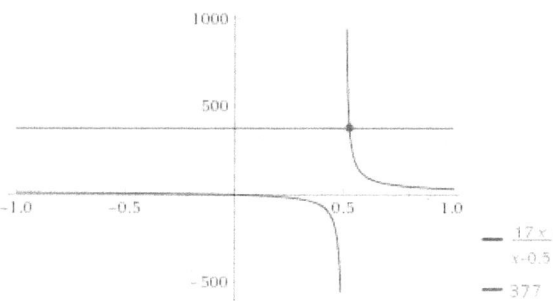

17*(x/(x-0.5))=377

Input:

$$17 \times \frac{x}{x - 0.5} = 377$$

Plot:

Alternate form:

$$\frac{17\,x}{x - 0.5} = 377$$

Solution:

$$x \approx 0.523611$$

The number 17 above in the image resulted in the drift by 0.5 degrees from the value of 12; the latter refers to the number of months per year. In other words, the significant cornerstone of linking the Impedance of Free Space with the Giza Plateau's design lies in the number 17 itself.[6]

CHAPTER REFERENCES

[1] https://en.wikipedia.org/wiki/Impedance_of_free_space
[2] Ibid.
[3] *The Calendar of Ancient Egypt: The Temporal Mechanics of the Giza Plateau (The Mill of Egypt) (Volume 1)*, Ibrahim Ibrahim.

---

[6] See previous chapters for important role which 17 plays in the equations.

12 THE SPHINX AND THE BEE
(This chapter is still in draft phase)

The king's title were represented by the Reed and the Bee as: the King of the South and the North (Nesu Bat[7]).[1] The word Bat, however, could have been derived from the word 'servant' as it is evident in the Arabic language.[2] Beeswax was used in the process of mummification and the word 'Mummy' was derived from the Arabic which means 'bitumen' in direct reference to, wax.[3] Bitumen was a major constituent for Third Intermediate period and Roman period Egyptian embalming techniques, in addition to traditional blends of pine resins, animal fats, and beeswax.[4]

The word 'Sphinx/Sphex' was originally derived from the word 'Bee', and the significance thereof lies in the words (Nesu Bat) which literally mean (Begotten and Resurrected). This observation of mine is confirmed by Exodus 2:3 where we read:

*When she could hide him no longer, she took for him a basket made of bulrushes and daubed it with **bitumen** and pitch. She put the child in it and placed it among the **reeds** by the river bank..*

This biblical text reveals to us that the original expression of the pharaonic title of southern and northern Egypt were coined as a

---

means for referring to the royal ritual of being begotten from the South and delivered by the Nile into the resurrection of the North; into the Land of the Bee i.e. Ta-Bitty [5] – which is the exact same word as: Tabernacle. What appears to me here is that ancient Egypt not only wanted to plagiarize the Israelite narrative of baby Moses, but was even performing the ritual of assigning that chosen role to every ancient Egyptian who acted like a bee (i.e. a servant) in serving the ultimate House, the House of Pharaoh.

The significance of the Sphinx lies in it serving as an Omphalos (i.e. Beehive). This populous nest serves as a ritualistic stage for enacting birth and resurrection twice a year; once at the time of each Equinox with half a beetle emerging from the Omphalos. The two events resemble the shell of peanuts, or more precisely: the Queen Bee Cups. These cups are oriented or protruding vertically from the face of the brood comb[6], just like the Sphinx were oriented along the West-East axis instead of the North-South one; the latter was the axis of the Nile which presumably served as the orientation of the brood comb.

Another significant evidence is seen on the Hedjet crown of Upper Egypt, which looks like an Omphalos or a Tabernacle being delivered from the Nile Upstream into Lower Egypt where the Deshret crown receives it and acts as a nest for it.

We know that the worker bees will only further build up the queen cup once the queen has laid an egg in it. As the young queen larva pupates with her head down, the workers cap the queen cell with beeswax. When ready to emerge, the virgin queen will chew a circular cut around the cap of her cell. Often the cap swings open when most of the cut is made, so as to appear like a hinged lid.[7]

The maximum number of days[8] in the Queen Bee's lifecycle before its emergence equals to 17.[8] Interestingly enough, this number is coupled with the number 7,000 as we see in Chapter 2. The reason behind such coupling lies in the following relation

$$11.459 \, RC = 6 \, meters := 6 \, days^{-1}$$

And

$$360 \, RC = (60 \times \pi) \, days^{-1}$$

So[9] if we were to look at the number 7,000 using the units of days, we get a metonic cycle[10] of 19 years and a conversion into RC units of *(Half the Width of the Great Pyramid's Base in meters)*^{2}. It is as if the metonic cycle were used to remove the conversion factor between RC and meter units.

The extra 700 days resemble the accumulation of 36 decans per year in a period of one metonic cycle. If we were, however, to look at the number 700 using the units of RC, we get the number of Earth rotations around the Sun per one tropical year.

The two equations above are very significant and reveal much to us. The ancient Egyptians did not only intend to build the

---

[8] Also refer to Chapter 3.

[9] For more details, refer to my book: *The Calendar of Ancient Egypt: The Temporal Mechanics of the Giza Plateau*.

[10] The metonic cycle is significant because it acts upon each solar year's amount of 10.88 days (in reference to the lunar calendar) and eventually adds up all these periods to an amount which equals the number of days per a lunar year minus the height of the Great Pyramid in meters.

pyramids and the other structures based on the 'biblical' number of days[11], they even went on to reciprocate a 'holy angle' out of that: i.e. $11.459°$. This angle is the one with which the pyramid of Menkaure was "misaligned" on the Giza Plateau. The equations above defines the relation between the six days and this angle by means of the Royal Cubit.[12]

$$\frac{11.459}{\pi} RC = \frac{6}{\pi} days$$

This definition was so holy to ancient Egyptians insomuch that it delivered to them, what I call, the **Primordial Square**:

$$3.6475 = (1\ Royal\ Cubit)^2$$
$$:= One\ Meter\ per\ Day$$

Interestingly enough, we also have

$$e^{3.6475} \approx \frac{Apex\ Angle\ of\ Great\ Pyramid}{2}$$

And even the number of decans is accounted for as we see here

$$354 = \varphi \times 60 \times 3.6475 = Days\ in\ a\ Lunar\ Year$$
$$= \frac{36^2}{3.6475} - 1$$

Also

$$(11.459 \times 3.6475)^2 \approx 440 \times 4$$

The Area of a circle with a radius of one meter (i.e. 1.9098 Royal Cubits) equals to the same amount by which the Pyramid of Menkaure is drifted away from the other pyramids on the Giza Plateau.

---

[11] Quotable: My Worldview.
[12] Which I asserted before and demonstrated how it was laid down as a measurement unit for time.

$$Area\ of\ a\ Circle = \pi r^2$$
$$11.459 = \pi \times (1.9098)^2$$

While

$$Area\ of\ a\ Section\ in\ a\ Circle = \frac{1}{2}\theta r^2$$

For an angle of 11.459° = 0.2 rad, we have the section area of a circle equals to a thousandth of 364.75 which is about a thousandth of a solar year.[13] It is now important to mention that it is circulated on the web that legends say that the Apis bull (whose name means "Bee" in Latin) produces 1000 bees as regenerated souls.[8][9] The bull in ancient Egypt marked the year and the thousand bees were separated throughout the year by an amount which equals to the natural logarithm constant.[14]

Also notice that[15]

$$10.88^2 = 32.5 \times 3.6475$$
$$(10.88\ RC)^2 = 32.5\ m/day$$

And in 70 days, we multiply the above equation with 70

$$(47.6)^2 = 32.5 \times 70\ m/day$$

Refer to Chapter 11 to understand the significance of the number 47. We also note that

$$(36 \times 0.5236)^2 = 355$$

And in radians, we can express this relation in the following:

---

[13] Note that $\pi/(6 \times 3.6475)$ also delivers a thousandth of the height of Khafre's pyramid in meters.

[14] For more information on the Engineering model of the section area of the circle which has been established on the Giza Plateau, please refer to Reference [11-3].

[15] For more information on this relation, please refer to [10].

$$(360 \times \underbrace{\sqrt{\frac{\pi}{180}}}_{\approx \frac{1}{7.5}})^2 = \frac{\pi}{180} \times 365 \times 355 \approx 47.5^2$$

CHAPTER REFERENCES

[1] *The Mummy: A Handbook of Egyptian Funerary Archaeology*, by Ernest A Wallis Budge.
[2] http://ribeekeeper.org/wp-content/uploads/2013/12/COMMENTARY-Bees-In-Ancient-Egypt.pdf
[3] Ibid.
[4] https://www.thoughtco.com/bitumen-history-of-black-goo-170085
[5] https://andrewgough.co.uk/articles_bee1/
[6] https://en.wikipedia.org/wiki/Queen_bee
[7] Ibid.
[8] *Paradigm Busters: Beyond Science, Lost History, Ancient Wisdom*, by J. Douglas Kenyon.
[9] *Ancient Code: Are You Ready for the Real 2012?*, by Various Authors.
[10] *32.5 System: The Complete Series Fused*, by Ibrahim Ibrahim.

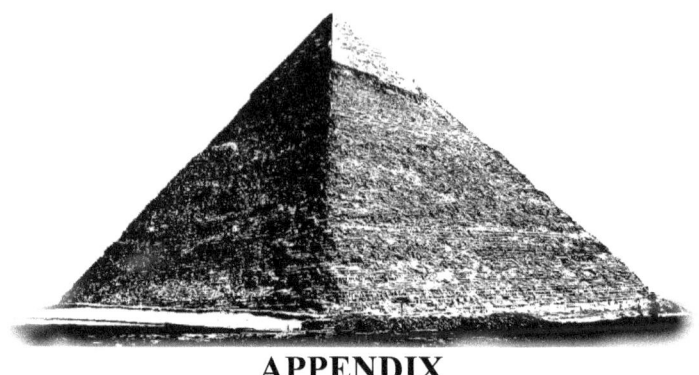

APPENDIX

I

ROBERT BAUVAL'S REMARKS ON THE SIGNIFICANCE OF USING THE MEASUREMENT UNITS OF METERS ON THE GPG

1. Height Over Perimeter Equals Radius Over Circumference

"Today we know that the circumference of a circle is calculated as the universal constant Pi times the Diameter of the circle. Now Pi is known to be 3.14159265359..... (to infinity). If you divide Pi by 6 you get 0.523598 which, as all mathematician and engineers will agree, can be rounded to 0.5236. Now this simple calculation will produce the same result whatever single unit of measure you use. However, it is only when you use the single meter unit that you get the exact value for the Royal Cubit used in the design of the Great Pyramid i.e. 3.14159 m/6 = 0.5236 m ... just to prick you interest further, if you take the base perimeter of the Great Pyramid, which is 1760 RC, and divide it by its height which is 280 RC, you get twice the value 6.285 RC. This is 2 x Pi. This would suggest that, mathematically speaking, the height of the pyramid to its base perimeter is the same as the radius of a circle to its circumference."

Robert Bauval
18 May 2016
[I-1]

2. The Pyramidion

"Egyptologists unanimously agree that the height of the Great Pyramid was originally 280 Royal Cubit (RC), and that the square base had sides of 440 RC. This means that the perimeter of the base is 440 x 4 = 1760 RC.
So far no arguments.
But here's the odd thing. If you divide 1760 RC into 12, you get 146.66. So far, all agree. No arguments.
But what does it mean?
Oddly, 146.66 is the height of the Great Pyramid in METERS.
But there is more.
The smallest possible "model" of the Great Pyramid in full Royal Cubits having a base divisible by 12 would be a 'pyramidion' having a base side of 3 RC and a height of 1.909 RC.
Oddly this converts to a pyramidion having a Height of 1 METER (100 cm) and a base side of 1.57 METER (157 cm.)
Did such a "model" exist?
Well, yes, perhaps."

Robert Bauval
18 May 2016
[I-2]

II
BESSEL FUNCTIONS TABLE

TABLE: Bessel functions of the first and second kind, for order 0 and 1

x	$J_0(x)$	$J_1(x)$	$Y_0(x)$	$Y_1(x)$
0.7	0.88120	0.32900	−0.19066	−1.10325
0.8	0.84629	0.36884	−0.08680	−0.97814
0.9	0.80752	0.40595	+0.00563	−0.87313
1.0	0.76520	0.44005	0.08826	−0.78121
1.1	0.71962	0.47090	0.16216	−0.69812
1.2	0.67113	0.49829	0.22808	−0.62114
1.3	0.62009	0.52202	0.28654	−0.54852
1.4	0.56686	0.54195	0.33790	−0.47915
1.5	0.51183	0.55794	0.38245	−0.41231
1.6	0.45540	0.56990	0.42043	−0.34758
1.7	0.39798	0.57777	0.45203	−0.28473
1.8	0.33999	0.58152	0.47743	−0.22366
1.9	0.28182	0.58116	0.49682	−0,16441
2.0	0.22389	0.57672	0.51038	−0.10703
2.1	0.16661	0.56829	0.51829	−0.05168
2.2	0.11036	0.55596	0.52078	+0.00149
2.3	0.05554	0.53987	0.51808	0.05228
2.4	0.00251	0.52019	0.51041	0.10049
2.5	−0.04838	0.49709	0.49807	0.14592
2.6	−0.09680	0.47082	0.48133	0.18836
2.7	−0.14245	0.44160	0.46050	0.22763
2.8	−0.18504	0.40971	0.43592	0.26355
2.9	−0.22431	0.37543	0.40791	0.29594

[II-1]

III
MAGIC SQUARES

Although they are/were used in esoteric topics instead of science because of the occult usage they were serving, but the numbers therein were allegedly – and yet possibly – linked to celestial phenomenon for each object in the *solar system*. I have demystified though the *magic square* of Mars in *32.5 System*. Further analytical observations I was able to embark upon were the following relations:

$$\frac{Magic\ Square\ of\ Mars\ (65)}{Magic\ Square\ of\ Jupiter\ (34)} = \frac{1}{0.523} = 1.9 \quad (1)$$

This is interesting indeed, since this is the relation between the units of meter and those of Royal Cubits. It might also be relevant to mention that

"According to Sitchin, Nibiru collided catastrophically with Tiamat (a goddess in the Babylonian creation myth the Enûma Eliš), which he considers to be another planet once located between Mars and Jupiter. This collision supposedly formed the planet Earth, the asteroid belt, and the comets. Sitchin states that when struck by one of planet Nibiru's moons, Tiamat split in two, and then on a second pass Nibiru itself struck the broken fragments and one half of Tiamat became the asteroid belt. The second half, struck again by one of Nibiru's moons, was pushed into a new orbit and became today's planet Earth."
[III-1]

Zecharia Sitchin mentions in his book series – *Earth Chronicles*, that there were a 12th planet that used to pass between Mars and

Jupiter. So what is significant here is to be aware that there is some kind of a message that is yet to be uncloaked and studied in regard to the relation between Jupiter and Mars that an observer on Earth is able to take notice of. Inspecting other relations of magic squares however gives some more interesting insights as well, for example

$$\frac{Magic\ Square\ of\ Sun\ (11)}{Magic\ Square\ of\ Mars\ (65)}$$
$$= 1 + \frac{1}{\sqrt{2}} = \frac{\sqrt{2}+1}{\sqrt{2}} \qquad (2)$$

And from *32.5 System*, we also have the celestial relations

$$\frac{1}{27.32} + \frac{1}{29.25} = \frac{1}{10\sqrt{2}} \qquad (3)$$

$$\frac{1}{27.32} - \frac{1}{29.25}$$
$$= \frac{1}{1{,}000(\sqrt{2}-1)} \qquad (4)$$

Rearranging the equations above gives

$$\rightarrow \frac{(2)}{(3)} = (4) \times 10{,}000 = \frac{100}{\pi + 1}$$

IV
THE NUMBER 309 FROM 32.5 SYSTEM

We have

$$\frac{50{,}000}{30{,}900} = 1.618$$

$$\frac{309}{50} = 6.18$$

And

$$\frac{354.37}{50} = 10 \times 0.7067 \approx 10\sqrt{0.5}$$
$$= \frac{10}{\sqrt{2}}$$

Since 309 refers to number of years, then we have an excess of about 9 years on the *solar calendar* drifting away from the lunar one. This happens on a yearly pace of

$$\frac{1}{9} = 0.1111 = 10.99 - 10.88$$

And more precisely

$$\frac{32.5}{300} = \frac{10 + 0.8333}{100} \approx \frac{10.8}{100}$$

THE ROLE OF 0.577

$$\frac{2\pi}{0.577} = 10.88 = 2\pi\sqrt{3}$$

$$\rightarrow \frac{2\pi \times 10.88}{0.577} = \frac{68.36}{0.577}$$

$$= 10.88^2 = 12\pi^2$$

Where

$$(1.577)^4 = 6.18$$

$$\frac{1}{1.577}^{16} = \frac{100,000}{68.3}$$

And

$$\pi - 1 = \frac{1}{0.683^2}$$

THE ROLE OF 0.57

$$\log_e \sqrt{\pi} \approx \frac{\sqrt{32.5}}{10}$$

THE ANGULAR LINK

More interesting is when we start handling the relations in units of degrees and link them to the celestial mechanics directly as shown below.

$$0.8333 \times 360 = 300$$

$$\frac{309}{0.8333} - 360 = 10.8$$

V
PRIME NUMBERS UP TO 1,000

There are

$$1.68 \times 100 = 168$$

primes in the first 1,000 numbers. [V-1]

It is interesting to see the order into which they were "arranged" so to speak and look for some information therein.

The List of Prime Numbers up to 1000
[V-2]

2, 3, 5, 7, 11, 13, 17, 19, 23, 29, 31, 37, 41, 43, 47, 53, 59, 61, 67, 71, 73, 79, 83, 89, 97, 101, 103, 107, 109, 113, 127, 131, 137, 139, 149, 151, 157, 163, 167, 173, 179, 181, 191, 193, 197, 199, 211, 223, 227, 229, 233, 239, 241, 251, 257, 263, 269, 271, 277, 281, 283, 293, 307, 311, 313, 317, 331, 337, 347, 349, 353, 359, 367, 373, 379, 383, 389, 397, 401, 409, 419, 421, 431, 433, 439, 443, 449, 457, 461, 463, 467, 479, 487, 491, 499, 503, 509, 521, 523, 541, 547, 557, 563, 569, 571, 577, 587, 593, 599, 601, 607, 613, 617, 619, 631, 641, 643, 647, 653, 659, 661, 673, 677, 683, 691, 701, 709, 719, 727, 733, 739, 743, 751, 757, 761, 769, 773, 787, 797, 809, 811, 821, 823, 827, 829, 839, 853, 857, 859, 863, 877, 881, 883, 887, 907, 911, 919, 929, 937, 941, 947, 953, 967, 971, 977, 983, 991, 997

An Interesting Number & Its Order

ORDER	NUMBER
12	37
33	137
37	157

We saw how 0.8333 produces a length of 10 when multiplied by 12 (<u>ten</u> multiples of 1.2) and how when further multiplied by 10, it refers to the celestial proportions

$$0.8333 = \frac{1}{1.2}$$

$$8.333 = \frac{100}{12}$$

since it takes about 8 minutes (<u>ten</u> multiples of 0.8333) for light to reach Earth from the Sun. With this information in hand, we were able to devise our way using Number Theory in tuning the proportions using these same numbers to produce

$$0.8333 = \frac{15 \; units}{18 \; units}$$

when magnifying the divine proportion of 1.618 by a factor of 100. This has been achieved by squaring the average value of 33 and by introducing the corner point of <u>ten</u> multiples of one 0.57[th]. Therefore, we see from the table above how the pattern is preserved for these numbers

$$(0.618 + 1) \times 100$$
$$(0.37 + 1) \times 100$$
$$(0.57 + 1) \times 100$$

And also note that

$$\frac{(0.618 \times 100) + 1,000}{2\pi}$$

$$= 1.6899 \times 100$$
$$\approx 168$$

LINKING 1.2732 WITH 0.83

I have presented in chapter 3 the significance of the number, 1.2732, and here it plays also a role in the following

$$(1 + 0.2732)^2$$
$$= 1 + \underbrace{2 \times 0.2732}_{(1+0.83)^{-1}}$$
$$+ 0.2732^2$$

INTRINSIC IMPEDANCE OF SPACE

$$0.8333 \times \frac{1{,}000\pi}{1{,}000\pi} = \frac{1{,}000\pi}{1{,}200\pi} = \frac{100\pi}{377\Omega}$$

VI
RESISTANCE AND DIRECTIVITY

Notice from Figures 9 (a & b) that 0.68 is implicitly included in the upper figure as well as explicitly in the lower one; and

$$60 \times \pi^2 = \frac{1,000}{1.68}$$

In other words, adding '1' to '0.68' gives it a role in the resistivity calculations alongside being relevant in finding out the directivity.

   Loop antennas with electrically small circumferences or perimeters have small radiation resistances that are usually smaller than their loss resistances. Thus they are very poor radiators, and they are seldom employed for transmission in radio communication. When they are used in any such application, it is usually in the receiving mode, such as in portable radios and pagers, where antenna efficiency is not as important as the signal-to-noise ratio. They are also used as probes for field measurements and as directional antennas for radiowave navigation. The field pattern of electrically small antennas of any shape (circular, elliptical, rectangular, square, etc.) is similar to that of an infinitesimal dipole with a null perpendicular to the plane of the loop and with its maximum along the plane of the loop. As the overall length of the loop increases and its circumference approaches one free-space wavelength, the maximum of the pattern shifts from the plane of the loop to the axis of the loop which is perpendicular to its plane. [VI-1]

VII
NASA TABLES

Here are the tables from NASA for the Earth-Sun distance (d) in astronomical units for Day of the Year (DOY):

DOY	d	DOY	d	DOY	d
1	0.98331	61	0.99108	121	1.00756
2	0.98330	62	0.99133	122	1.00781
3	0.98330	63	0.99158	123	1.00806
4	0.98330	64	0.99183	124	1.00831
5	0.98330	65	0.99208	125	1.00856
6	0.98332	66	0.99234	126	1.00880
7	0.98333	67	0.99260	127	1.00904
8	0.98335	68	0.99286	128	1.00928
9	0.98338	69	0.99312	129	1.00952
10	0.98341	70	0.99339	130	1.00975
11	0.98345	71	0.99365	131	1.00998
12	0.98349	72	0.99392	132	1.01020
13	0.98354	73	0.99419	133	1.01043
14	0.98359	74	0.99446	134	1.01065
15	0.98365	75	0.99474	135	1.01087
16	0.98371	76	0.99501	136	1.01108
17	0.98378	77	0.99529	137	1.01129
18	0.98385	78	0.99556	138	1.01150
19	0.98393	79	0.99584	139	1.01170
20	0.98401	80	0.99612	140	1.01191
21	0.98410	81	0.99640	141	1.01210
22	0.98419	82	0.99669	142	1.01230
23	0.98428	83	0.99697	143	1.01249
24	0.98439	84	0.99725	144	1.01267
25	0.98449	85	0.99754	145	1.01286
26	0.98460	86	0.99782	146	1.01304
27	0.98472	87	0.99811	147	1.01321
28	0.98484	88	0.99840	148	1.01338

29	0.98496	89	0.99868	149	1.01355
30	0.98509	90	0.99897	150	1.01371
31	0.98523	91	0.99926	151	1.01387
32	0.98536	92	0.99954	152	1.01403
33	0.98551	93	0.99983	153	1.01418
34	0.98565	94	1.00012	154	1.01433
35	0.98580	95	1.00041	155	1.01447
36	0.98596	96	1.00069	156	1.01461
37	0.98612	97	1.00098	157	1.01475
38	0.98628	98	1.00127	158	1.01488
39	0.98645	99	1.00155	159	1.01500
40	0.98662	100	1.00184	160	1.01513
41	0.98680	101	1.00212	161	1.01524
42	0.98698	102	1.00240	162	1.01536
43	0.98717	103	1.00269	163	1.01547
44	0.98735	104	1.00297	164	1.01557
45	0.98755	105	1.00325	165	1.01567
46	0.98774	106	1.00353	166	1.01577
47	0.98794	107	1.00381	167	1.01586
48	0.98814	108	1.00409	168	1.01595
49	0.98835	109	1.00437	169	1.01603
50	0.98856	110	1.00464	170	1.01610
51	0.98877	111	1.00492	171	1.01618
52	0.98899	112	1.00519	172	1.01625
53	0.98921	113	1.00546	173	1.01631
54	0.98944	114	1.00573	174	1.01637
55	0.98966	115	1.00600	175	1.01642
56	0.98989	116	1.00626	176	1.01647
57	0.99012	117	1.00653	177	1.01652
58	0.99036	118	1.00679	178	1.01656
59	0.99060	119	1.00705	179	1.01659
60	0.99084	120	1.00731	180	1.01662

DOY	d	DOY	d	DOY	d
181	1.01665	241	1.00992	301	0.99359
182	1.01667	242	1.00969	302	0.99332
183	1.01668	243	1.00946	303	0.99306

184	1.01670	244	1.00922	304	0.99279
185	1.01670	245	1.00898	305	0.99253
186	1.01670	246	1.00874	306	0.99228
187	1.01670	247	1.00850	307	0.99202
188	1.01669	248	1.00825	308	0.99177
189	1.01668	249	1.00800	309	0.99152
190	1.01666	250	1.00775	310	0.99127
191	1.01664	251	1.00750	311	0.99102
192	1.01661	252	1.00724	312	0.99078
193	1.01658	253	1.00698	313	0.99054
194	1.01655	254	1.00672	314	0.99030
195	1.01650	255	1.00646	315	0.99007
196	1.01646	256	1.00620	316	0.98983
197	1.01641	257	1.00593	317	0.98961
198	1.01635	258	1.00566	318	0.98938
199	1.01629	259	1.00539	319	0.98916
200	1.01623	260	1.00512	320	0.98894
201	1.01616	261	1.00485	321	0.98872
202	1.01609	262	1.00457	322	0.98851
203	1.01601	263	1.00430	323	0.98830
204	1.01592	264	1.00402	324	0.98809
205	1.01584	265	1.00374	325	0.98789
206	1.01575	266	1.00346	326	0.98769
207	1.01565	267	1.00318	327	0.98750
208	1.01555	268	1.00290	328	0.98731
209	1.01544	269	1.00262	329	0.98712
210	1.01533	270	1.00234	330	0.98694
211	1.01522	271	1.00205	331	0.98676
212	1.01510	272	1.00177	332	0.98658
213	1.01497	273	1.00148	333	0.98641
214	1.01485	274	1.00119	334	0.98624
215	1.01471	275	1.00091	335	0.98608
216	1.01458	276	1.00062	336	0.98592
217	1.01444	277	1.00033	337	0.98577
218	1.01429	278	1.00005	338	0.98562
219	1.01414	279	0.99976	339	0.98547
220	1.01399	280	0.99947	340	0.98533

221	1.01383	281	0.99918	341	0.98519
222	1.01367	282	0.99890	342	0.98506
223	1.01351	283	0.99861	343	0.98493
224	1.01334	284	0.99832	344	0.98481
225	1.01317	285	0.99804	345	0.98469
226	1.01299	286	0.99775	346	0.98457
227	1.01281	287	0.99747	347	0.98446
228	1.01263	288	0.99718	348	0.98436
229	1.01244	289	0.99690	349	0.98426
230	1.01225	290	0.99662	350	0.98416
231	1.01205	291	0.99634	351	0.98407
232	1.01186	292	0.99605	352	0.98399
233	1.01165	293	0.99577	353	0.98391
234	1.01145	294	0.99550	354	0.98383
235	1.01124	295	0.99522	355	0.98376
236	1.01103	296	0.99494	356	0.98370
237	1.01081	297	0.99467	357	0.98363
238	1.01060	298	0.99440	358	0.98358
239	1.01037	299	0.99412	359	0.98353
240	1.01015	300	0.99385	360	0.98348
				361	0.98344
				362	0.98340
				363	0.98337
				364	0.98335
				365	0.98333
				366	0.98331

Link: http://landsathandbook.gsfc.nasa.gov/excel_docs/d.xls

APPENDIX REFERENCES

[I-1]
https://www.facebook.com/photo.php?fbid=10153694156049016

[I-2]
https://www.facebook.com/robbo50333/posts/10153694725024016

[II-1]
http://onlinelibrary.wiley.com/doi/10.1002/9781118093184.app4/pdf

[III-1] https://en.wikipedia.org/wiki/Zecharia_Sitchin

[V-1] *The Music of the Primes: Searching to Solve the Greatest Mystery in Mathematics*, by Marcus du Sautoy (p. 90).
[V-2] http://www.miniwebtool.com/list-of-prime-numbers/?to=1000

[VI-1] See reference [13].

INDEX

And when a sign comes to them, they say, "Never will we believe until we are given like that which was given to the messengers of God." God knows best **where** He places His message. There will afflict those who committed crimes debasement before God and severe punishment for what they used to conspire.

~ Quran 6:124

32.5

If 32.5 were just a number rather than a complete System, I wouldn't have received thereby the keys for unlocking some of the mysteries of the Great Pyramid of Giza.

ABOUT THE AUTHOR

Ibrahim researches and studies the scientific theories and their applications, especially in Physics. He is engaged in the process of developing frameworks, onto which our theories are built, to enhance the already available layers of presentation while constructing new ones when needed and to further facilitate access unto wide range of interconnected resources of information and scientific facts; Ibrahim's work contributes in raising public awareness of significantly factual and trustworthy theories and observations.

He is also an Engineer by Profession, a Physicist by Academic Achievement and in general a Researcher yet on the Scientific Method. In his spare time, he navigates into other scientific fields using MOOCs to broaden his spectrum of knowledge. He became in 2015 a Community Teaching Assistant for both of the MOOCs, 'Genomic Data Science with Galaxy', and, 'Introduction to Genomic Technologies', of Johns Hopkins University. https://twitter.com/Ibrahim_Squared